Report on the Variations of the Magnetic Intensity observed at different Points of the Earth's Surface. By Major EDWARD SABINE, *R.A., F.R.S.*]

[With Plates.]

It has been justly remarked by M. de Humboldt, "that the phænomena of the earth's magnetism, in its three forms of variation, dip, and intensity, have of late years been examined with great care, in the most different zones, by the united efforts of many travellers; and that there is scarcely any branch of the physical knowledge of the earth in which, in so small a number of years, so much has been gained towards an acquaintance with its laws, though not perhaps with its causes." (*Ann. der Physik*, vol. xv. p. 320.)

Be it here remarked, that it is to the example and the writings of this illustrious philosopher that the accelerated progress in this, as in so many other branches of physical science, is eminently due. His writings exhibit, in the most pleasing manner, the delightful, the never-failing interest which such pursuits afford, awaken thereby a taste for them in those who were previously unconscious of its existence, and stimulate its exercise in all. It is in this respect that M. de Humboldt has been not only a great promoter of science, but a moral benefactor to many; for it is the privilege of such pursuits that tedious hours are little known to the mind that engages in them, and the enjoyment which they yield is unimpaired by advancing years*.

M. de Humboldt's remark is particularly true in regard to the magnetic intensity. At the commencement of the present cen-

---

* The surviving friends of the late Major Rennell have, in their recollection of that true philosopher, when engaged in his latter years in his important work on the currents of the Atlantic Ocean, a memorable example of this power of physical research, to preserve its interest vivid and unbroken amidst the infirmities of declining years.

tury, the bare fact of there being any difference whatsoever in the intensity of the magnetic force in different parts of the earth was unattested by a single published observation. The maps attached to this memoir exhibit the progress which investigation has made in the years that have since elapsed. They contain 753 distinct determinations, at 670 stations widely distributed over the earth's surface; leaving, it is true, much still to be desired;—but in what has been accomplished, leading to conclusions so remarkable, in regard to the phænomena of magnetism, on the largest scale presented to us by nature, as to stimulate greatly to more extensive research.

I have sought to embody in this report on the variations of the magnetic intensity, all the materials which have been obtained by the labours of observers of all nations, in all parts of the world;—to present them in the form best fitted to add to our knowledge;—and to call attention to the general conclusions, to which we are conducted by an attentive consideration of the facts of observation, when thus brought together in one view. A large portion of these determinations are here published for the first time. The observations of Capt. de Freycinet, Capt. King, Mr. Douglas, Capt. Fitz Roy, Capt. Ross, and Major Estcourt are wholly new, the original observations having been recently communicated to me by the respective observers, and calculated and arranged by me. Messrs. Hansteen and Due's Siberian observations, and M. Erman's in the Pacific and Atlantic oceans, have been furnished to me by the liberality of those gentlemen, calculated as they appear here. Of the results previously published, the greater number are collected from different foreign works which have little circulation in this country; and some of these, as well as a part of my own observations published in this country several years ago, have required additional calculations, for the purpose of bringing them into the general comparison.

I have divided the report into three sections; the first, containing a condensed historical notice of each of the several series of observations, by which our knowledge of the magnetic intensity has been progressively advanced; the second, comprising the whole of the results, classed according to the values of the intensity, and arranged in a tabular form; and the third, containing a summary of the principal general conclusions in regard to the system of terrestrial magnetism, which are deducible from the facts thus collected.

I have endeavoured to confine the historical notices in the first section within the narrowest limits compatible with the primary object, that of including in each notice all the circumstances required to be known in order to estimate rightly the value of the results. In the case of observations which are either wholly or partly new, these particulars are not to be found elsewhere; and in the case of those series, the published accounts of which are contained in foreign works rarely met with in this country, it has appeared desirable,—whilst giving every direction which may facilitate a reference to the original publication,—to make the account here given complete in all particulars essential to a just estimation of the value of the results, independently of such reference. The details necessary for this purpose may render this portion of the report occasionally tedious to the general reader, who will be principally interested by that section which contains the general conclusions.

SECTION I.—HISTORICAL NOTICES.

It is to France we owe the first rightly directed experimental inquiry on this subject. The instructions, drawn up by the members of the French Academy of Sciences for the expedition of La Perouse, contain a recommendation that the time of vibration of a dipping needle should be observed at stations widely remote, as a test of the equality or difference of the magnetic intensity; suggesting also with a sagacity anticipating the result, that such observations should particularly be made at those parts of the earth where the dip was greatest and where it was least.

The experiments, whatever their results may have been, which in compliance with this recommendation were made in the expedition of La Perouse, perished in its general catastrophe; but the instructions survived, and bore fruit in the earliest recorded observations of the variations of the magnetic intensity, which are those published by M. de Rossel in the second volume of the *Voyage de Dentrecasteaux* in search of La Perouse.

Rossel, 1791-1794.—These observations, though made in the years above-mentioned, were not published until 1808. They were made with a needle vibrated in a dip circle of Le Noir, coming to rest disadvantageously soon for the purpose of experiments on the intensity. The needle continued in vibration little more than three minutes; consequently incidental errors would bear a very large proportion to the total time of vibration; a disadvantage which appears to have been in a great degree counteracted by the very great care bestowed on the observation. The needle was vibrated at Brest in 1791, before the voyage commenced; and, successively, at Teneriffe; Van Diemen's Land, in May 1792; at Amboyna, in October of the same year; again at Van Diemen's Land, in February 1793; and at Surabaya in Java, in 1794. With this last observation the published results terminate; there is no record of the vibrations having been repeated on the return to France, for the purpose of testing the constancy of the magnetism of the needle, a step which subsequent experience has shown to be most important. The connexion of all the foreign stations with Europe is consequently imperfect; and the values of the intensity at those stations, relatively to any standard value in Europe, could only be computed, subject to the uncertainty arising from the possibility of a change in the magnetic condition of the needle. The conclusion drawn by M. de Rossel, of the increase of the intensity in receding from the equatorial to the higher latitudes, was, however, fully borne out and substantiated, in regard to the southern hemisphere, by the observations at Van Diemen's Land in 1792 and 1793, compared with the intermediate vibrations at Amboyna. These form a comparison complete in all respects, and to the certainty of which nothing is wanting. It is independent of any change the needle may have undergone before or afterwards; the correspondence of the time of vibration at Van Diemen's Land in May 1792 and February 1793, proving the needle to have been steady in that interval. The increase in the intensity between Amboyna and Van Diemen's Land was in the proportion of 1 to 1·67, a difference far too great to be attributed to any supposable errors or accidents of observation. It is this determination which unquestionably entitles Admiral de Rossel to the distinction which he has always enjoyed, of having been the first who ascertained that the magnetic intensity is different at different positions on the earth's surface: although his observations were not published until after those of M. de Humboldt in 1798–1803, by which the same fact was more largely established.

As M. de Rossel's observations have not, I believe, been published in any English work, I have subjoined a table containing an abstract of all their essential particulars.

Station.	Date.	Lat.*	Long.*	Dip.	Time of Vibration.
Brest	20 Sept., 1791	48 24	355 34	71 30 N.	2·02
Teneriffe	21 Oct., 1791	28 28	343 42	62 25 N.	2·081
Van Diemen's Land	11 May, 1792	43 32 S.	146 57	70 50 S.	1·869
Amboyna	9 Oct., 1792	3 42 S.	128 08	20 37 S.	2·403
Van Diemen's Land	7 Feb., 1793	43 34 S.	146 57	72 22 S.	1·850
Surabaya	9 May, 1794	7 14 S.	112 42	25 20 S.	2·429

The times of vibration are in infinitely small arcs, being reduced by M. de Rossel, by means of a table which accompanies the observations in the original publication.

M. de Rossel's observations at Van Diemen's Land were made at a port on the S.E. part of the island. Capt. Fitz Roy has recently determined the value of the intensity at Hobart Town, about 40 miles north of M. de Rossel's station, to be 1·817, in terms of a comparative scale in general use adopted in this Report, of which an explanation will be given in the sequel. Suffice it at present to say, that in the same scale the force at Paris = 1·348, and at London 1·372. Capt. Fitz Roy's observations will be found in their place in the course of this Report. If we take his value of the intensity at Hobart Town for the force at M. de Rossel's station, we have 1·097 as the force at Amboyna. By means of Capt. Fitz Roy's observation at Van Diemen's Land, I have been thus enabled to connect M. de Rossel's determination at Amboyna with Europe, and it is accordingly entered in the general table.

Humboldt, 1798–1803.—These observations were made in the course of M. de Humboldt's well-known journey to equinoctial America. Various partial notices of them have appeared at different times and in different works, but a complete account, communicated by M. de Humboldt himself, may be found in the xvth volume of the *Annalen der Physik*, from which the results employed in this memoir are derived. The observations were made with a dipping needle of Le Noir, selected by M. Borda. It vibrated considerably longer before coming to rest than the needle employed by M. de Rossel, so as to allow the number of vibrations performed in ten minutes to be taken as the measure of the intensity at the different stations. The time of vibration at Paris was observed in October 1798, be-

* All the longitudes in this Report are east of Greenwich, unless otherwise expressed; and all the latitudes are north unless they are designated otherwise.

fore M. de Humboldt's departure; but as the needle was left in Mexico, those observations could not be made on the return to Europe, by which its magnetic invariability might have been assured. The circumstances are greatly to be regretted, whatever they may have been, which deprived a suite of observations so extensive, and on which so much care and labour had been bestowed, of a final confirmation, which can hardly be supplied in an equally satisfactory degree by any less direct evidence. Fortunately, indirect means are not altogether wanting in this case, and we may infer from them that up to the beginning of 1800 M. de Humboldt's needle had undergone no change; and that if subsequently to that date it lost magnetism, the alteration was not considerable. The observations in Paris were made in 1798. Between August 1799 and February 1800, M. de Humboldt made thirteen determinations of the intensity on the Spanish main, between the latitudes of 10° and 11°, and the longitudes of 292½ and 296¼. The mean of these is an intensity of 1·196. In 1822 the value of the intensity at Trinidad, in lat. 10° 39' and long. 298½, was determined, by observations made by myself (to be discussed hereafter), to be 1·204. The result of this comparison is extremely satisfactory; and being derived, on M. de Humboldt's side, from observations with one needle at several stations, and on mine from several needles at one station, a fair conclusion may be drawn, that in the beginning of 1800 his needle retained its magnetism unimpaired. In January, 1801, M. de Humboldt's needle gave for the intensity at Havannah 1·359; mine, in 1822, 1·499. In this comparison the agreement is less perfect; there is a greater difference than is usual between the results of different observers at the same station; and it is such as would be occasioned by a loss of magnetism in M. de Humboldt's needle, but not to an amount that would impair in a material degree the value of his important series. Against any precise inference, however, to be drawn from these comparisons, there is, 1st, the difference of the dates at which the respective intensities were determined; 2nd, a small difference in longitude of the localities of the first comparison; and 3rd, those circumstances of a local and instrumental nature which must affect every such comparison.

In the account which M. de Humboldt has given of his observations there is no mention made of corrections having been applied for the arcs of vibration or for the temperature of the needle; but in such an extensive series, corrections on these accounts are of minor importance.

The number of land-stations at which the intensity was observed appears to have been 77, all of which are entered in the general table in this memoir.

Besides the land-stations, there are 12 geographical positions, in which M. de Humboldt observed the vibrations of the needle on board ship. There are two great and obvious disadvantages in such observations, compared with those on land, viz. the motion, and the iron, of the vessel. On the other side should be noticed, the space interposed between the instrument and the solid materials of the earth's surface, many of which are known to exercise a very considerable disturbing influence on the needle. As opinions may, and I believe do, vary in regard to the degree of relative value to be allowed to observations of intensity made at sea and on land, and as it is not a point on which, from personal experience, I feel qualified to decide, I have placed the sea-observations in a separate table, and subjoin them here.

Latitude.	Longitude.	Date.	Intensity.
38 52	345 59	1799	1·315
37 26	345 49	1799	1·315
34 30	345 26	1799	1·230
31 46	345 17	1799	1·261
24 53	341 23	1799	1·283
3 02 S.	279 54	1803	1·067
21 29	336 39	1799	1·261• } 1·256
19 54	333 36	1799	1·251• }
14 15	314 18	1799	1·283• } 1·256
13 02	309 23	1799	1·230• }
10 46	301 27	1799	1·178• } 1·220
11 01	297 30	1799	1·261• }

The results marked with an asterisk were observed on the passage across the Atlantic, between Teneriffe and Trinidad, a part of the ocean where no land exists, and where, consequently, the results obtained at sea furnish the only attainable evidence. On examination, they present differences among themselves considerably greater than is usual in land results ; but by combining them in pairs, as shown in the table, and using the mean latitude, longitude, and intensity of each pair, these partial differences greatly disappear. I have entered the mean latitude, longitude, and intensity of these three pairs in the general table.

Humboldt and Gay Lussac, 1805–1806.—These observations

were made during a tour in France, Switzerland, Italy, and Germany, with a needle suspended by fibres of silk, vibrating in the plane of the horizon, and measuring the horizontal component of the magnetic intensity. The dip was observed at the same time with a dipping-needle of Lenoir (the same that had been used in the *Voyage de Dentrecasteaux*), supplying the means of computing the total intensity from its horizontal component. An account of these observations was published by M. Gay Lussac in the 1st volume of the *Memoires de la Société d'Arcueil*. The values of the intensity were given in reference to the force at Paris, where the needle was vibrated at the close of the series, but not at its commencement. M. Gay Lussac infers that no change took place in the magnetism of the needle throughout the series, from its having had the same time of vibration at Milan on two occasions, viz. in going and in returning, at six months' interval. As no dates are given, the stations at which the strict comparability of the force was thereby ensured can only be conjectured. It is probable that no corrections were applied either for the arcs or for differences of temperature, as neither of these circumstances is noted in the record. The number of stations of known geographical position is 19, 16 of which are inserted in the general table in this memoir. The other stations were in the crater, on the side, and at the foot of Vesuvius, where the results were considered by the observers to be affected, as no doubt they were, by the proximity of the lava.

Sabine, 1818, 1819, 1820.—These observations were made in the first and second voyages of northern discovery to Baffin's Bay and the Polar Sea. Aware of the magnetic importance of the regions to be explored, and anxious duly to improve such opportunities, I sought diligently to provide myself with instruments adequate to the occasion. Those furnished by Government were by no means so; but it fortunately happened that my brother-in-law Mr. Browne possessed and entrusted to me a dip circle and needle of very superior character, made by Nairne and Blunt, and similar in all respects to the one made under Mr. Cavendish's directions, and described by him in the 66th vol. of the *Phil. Trans.* The needle vibrated about eight minutes before coming to rest; and probably, from its age, had long acquired the state of steady magnetism which it was proved to possess during these voyages, its time of vibration being almost identical when observed in London in March, 1818, in March, 1819, and in December, 1820*.

* The observations of March, 1819, and December, 1820, are recorded in

The observations of the voyage of 1818 were published in the *Phil. Trans.* for 1819; those of the voyage of 1819-20, partly in the appendix to the narrative of that voyage, and partly in my work entitled *Pendulum and other Experiments,* published in 1825. In these publications the results were deduced without any corrections having been made for the arc of vibration or the temperature of the needle. On this occasion I have introduced both these corrections. That for the arc has been computed by means of the table published in the *Voyage de Dentrecasteaux,* which I find to reduce the vibrations in the different arcs so nearly to an equality as fully to justify its employment. The arcs themselves are stated in the printed record of the observations. The temperatures on the different days of observation are taken from the record of the external thermometer in the Meteorological Journal, and the corrections are computed by the usual formula for that purpose, in which the coefficient ·0004 has been determined by experiments with the same needle in high and low temperatures.

In the voyage of 1819-1820 I furnished myself, besides the dipping-needle, with three horizontal needles, and an apparatus for their vibration. These would have been of great use had it been our good fortune to have returned to Europe by the way of the Pacific; but the method of deducing the total intensity by means of horizontal needles almost ceases to be available in countries where the dip so nearly approaches 90°, and where small incidental errors in the determination of the dip will so greatly affect the conclusion as to the force. Accordingly, I have at no time brought the observations with the horizontal needles in this voyage in comparison with the results given by the dipping-needle. There is, however, an incidental purpose of some value which they may serve, which did not occur to me when the record of the observations was printed, and which is worth noticing, as it may be useful on similar occasions, should there be such. The horizontal vibrations, though inappropriate in such circumstances to furnish the total intensities, give as correct measures of the relative values of the horizontal component

the Appendix of the second Polar Expedition. From the circumstance of the narrative and appendix of that voyage having been published at an interval of some months apart, the copy of the narrative which reached M. Hansteen was unaccompanied by the appendix, which it seems he has never seen. The abstract of the results, published in another work from whence he has taken them, refers to the full record of the observations in the appendix, and omits their dates, and Mr. Hansteen has consequently been at a loss to know whether the vibrations were observed both before and after the voyage of 1819—1820. By consulting the original account, he will see that this necessary care was not omitted.

of the force at any two stations, as the vibrations of the dipping-needle do of the total force. If, then, T is the time of horizontal vibration, and D the dip at a primary station, where the total force is taken as unity,—and if T' and D' are the same quantities at another station, where I' is the value of the total intensity derived by the vibrations of the dipping-needle,—

$$\cos D' = \frac{T^2 \cdot \cos D}{I' \, T'^2};$$ and we thus get a determination of the

dip distinct from the ordinary method, and independent of the instrumental errors from which it is so difficult to clear the dipping-needle, especially one in which the poles are not reversed in every observation.

Employing the observations at Melville Island, printed in the appendix to the account of that voyage, in this manner, we obtain the dip at Melville Island by the three horizontal needles as follows, viz.

Needle 1 . . . . 88° 44'
Needle 2 . . . . 88 46
Needle 3 . . . . 88 48

The direct observation by the dipping-needle was 88° 43'·5.

The following table exhibits the results of the observations of intensity in the two north polar voyages above noticed, corrected for temperature and arc, and expressed in terms of the general scale.

Station.	Latitude.	Long.	Therm.	Time of Vibration.		Intensity.
				Observed.	Correct.	
	° '	° '				
London, 1818	⎫			⎧ 480	472·0	⎫
London, 1819	⎬ 51 31	359 52	48	⎨ 482	473·5	⎬ 1·372
London, 1820	⎭			⎩ 480	472·9	⎭
Shetland, 1818	60 09	358 48	44	470	461·7	1·434
On Ice	68 22	306 10	34	440	432·1	1·643
Hare Island	70 26	305 08	34	443	434·9	1·622
On Ice	75 05	299 37	33	447·2	439·4	1·590
On Ice	75 51	296 54	33	443·6	435·6	1·618
On Ice ,..................	76 45	284 00	33	435·0	429·1	1·666
On Ice	76 08	281 39	33	436·0	430·0	1·659
On Ice	70 35	293 05	33	436·0	429·7	1·661
On Ice, 1819............	64 00	298 10	32	437·4	435·0	1·621
Possession Bay	73 31	282 38	40	439·5	432·9	1·637
Regent's Inlet	72 45	270 19	32	439·0	428·9	1·668
Byam Martin's Island .	75 10	256 16	32	442·5	430·7	1·653
Melville Island	74 27	248 18	20	444·3	434·6	1·624
Winter Harbour	74 47	249 12	43	446·2	432·6	1·638

Hansteen, 1819–1825.—In 1819 M. Hansteen, having completed and published his elaborate exposition of the theory of the earth's magnetism, to which he had been conducted by the study of the phænomena of the variation and dip as far as they were then known, entered into the field of experimental research, in which he has since rendered such important practical services to his favourite science. His exceedingly portable apparatus for determining the intensity by horizontal needles is too well known to need description here ; and his good fortune in possessing a needle of remarkably steady magnetism, supplied by Mr. Dollond, renders little more necessary to be said in regard to his determinations, than to refer to the publications in which they may be found, and to enter them in the general table. From 1819 to 1824 his observations were confined to Norway and the shores of the Baltic, and were published in the iiird vol. of the *Ann. der Physik*, the intensity stations being 37. In 1825 he extended them round the shores of the Gulf of Bothnia; and the determinations of that year, being 30 in number, were published, first, in the ixth vol. of the *Ann. der Physik*, and, secondly, with corrections, in the *Astro. Nach.*, No. 146.

Erichsen, 1824; *Keilhau and Boeck*, 1825–1827; *Erman*, 1826.—I have classed these observations together, because they were all made, I believe, at the instance and with the apparatus of M. Hansteen, and were communicated to the public through him in the *Astro. Nach.*, No. 146. Captain Erichsen's consist of 3 stations on the shores of the Baltic, and in Germany ; Messrs. Keilhau and Boeck's of 9 stations in Germany; and M. Erman's of 2 stations in Germany. They were all connected with Paris through Christiania, and are entered in the general table.

Sabine, 1822–1823.—These observations were made during two voyages, in which I was furnished by the British Government with a vessel for my conveyance to stations at remote latitudes from each other, for the purpose of determining the amount of the ellipticity of the earth by means of the pendulum. The first voyage was to the equatorial shores of the African and American continents, and the second to the north of Europe, Greenland, and Spitzbergen. For these voyages I supplied myself with as many as six horizontal needles, in anticipation that some amongst them might prove unsteady in their magnetism. The observations with all the needles, and at all the stations visited, were published in 1825, with the account of the pendulum experiments.

One of the needles, No. 2, lost so much of its magnetism in

the first voyage that it was not used in the second.   Another,
No. 1, appears to have been subject to fluctuations in its
magnetic condition, rather than to have undergone permanent
or uniform gain or loss.   M. Hansteen, who has discussed
these observations at some length in the ixth volume of
the *Annalen der Physik*, has rejected the results with these
two needles whenever they differed considerably from those
of the other four; but has retained and allowed weight
in the general mean to such of their results as appeared
to agree with the other needles.   Nos. 3, 4, 5, and 6 showed
on their return to England small and comparatively unim-
portant differences from their times of vibration previous to
their departure.   M. Hansteen has applied corrections on this
account to the intervening observations, according to their
dates.   One of my stations having been Drontheim in Norway,
which was visited by M. Hansteen himself for the same purpose
in 1825, two years after I had been there, it became a station
common to our respective series ; and he was thereby enabled
to compute the values of the intensity at all the stations visited
by me, relatively to the force at Drontheim, which he had
already compared with Paris by observations at Drontheim
and Christiania, and at Christiania and Paris.   The values so
computed and published by M. Hansteen in the volume of the
Ann. der Physik referred to, are here subjoined, for the purpose of exhibiting them in comparison with my own deductions.   The latter are made from the observations with Nos. 3,
4, 5, and 6 alone, those of Nos. 1 and 2 being put wholly
aside.   The times of vibration of each needle at the different
stations, as originally published in 1825, have received three
corrections : one, when necessary, for change of magnetism,
assigned on the principle of uniform gain or loss ; a second,
to diminish the observed times of vibration to the corresponding times in infinitely small arcs ; and a third for reduction to
a standard temperature of the needle, the coefficients for
the formula having been determined experimentally for each
needle.   The values of the intensity in my deductions are given
relatively to the force in Paris, by my own comparison of the
force in London and in Paris, which will be noticed hereafter.
There are, therefore, several particulars in which M. Hansteen's mode of deduction and mine differ ; but it is interesting
to perceive how nearly the results agree.   The values calculated
by M. Hansteen are almost everywhere slightly in defect of those
computed by me.   This arises from the force at Drontheim being somewhat less by M. Hansteen's observations than by mine ;
and as he has compared the intensity at all my stations with that

at Paris through the observations at Drontheim, the original
difference between us at Drontheim pervades the whole series.

Place.	Hansteen.	Sabine.	Place.	Hansteen.	Sabine.
Bahia	0·894	0·898	Madeira	1·382	1·373
Ascension	0·900	0·920	Jamaica	1·414	1·436
St. Thomas	0·921	0·931	Drontheim	1·430	1·442
Maranham	1·006	1·016	Grand Cayman .	1·430	1·454
Sierra Leone ...	1·043	1·053	Havanna.........	1·493	1·499
Gambia River...	1·129	1·141	Hammerfest ...	1·493	1·506
Port Praya	1·184	1·193	Greenland	1·512	1·530
Trinidad	1·183	1·204	Spitzbergen	1·531	1·562
Teneriffe	1·300	1·313	New York	1·794	1 803

In the deductions contained in this table (both in M. Hansteen's and mine) the dips employed are those which M. Hansteen has calculated from my published observations. They
differ occasionally a minute or two from my calculated results,
but in no instance does the difference amount to 3′.

Lütke, 1826–1829.—These observations were made by
Captain (since Admiral) Lütke, of the Russian Imperial Navy,
in a voyage of circumnavigation in H.I.M. ship Siniavin. At the
request of Capt. Lütke, M. Lenz, of the Imperial Academy of
Sciences at St. Petersburg, undertook to arrange them for
publication, and they have since been published in the German
language in the Memoirs of the Imp. Acad. of Sciences for 1835.
I was indebted to the friendship of Capt. Lütke for an early
knowledge of these observations, having received a copy of them
in a letter from Norfolk Sound in July 1827; but the present
notice, as well as the results entered in the table, are taken
from the published account.

M. Lenz's memoir is divided into two sections,—on the observations of Dip,—and on those of Intensity. Our present
purpose is with the latter section.

The observations of intensity were made with one dipping and
five horizontal needles. The dipping-needle was 3½ inches in
length, with a steel axle, and was reserved exclusively for measuring the intensity by its vibrations, as there were two other
dipping-needles for observations of the dip. The horizontal
needles were of various shapes, cylindrical, rhomboidal, and
elliptical, but all of the same length, i. e. two English inches.
They were obtained in England when the Siniavin was on her
outward passage. The apparatus in which they were to have

been used was unfortunately broken in pieces in the carriage from London to Portsmouth by mail. It had been Capt. Lütke's intention to have vibrated the needles at Portsmouth before his departure, and again at the same spot on his return from the Pacific; so that all the observations of his voyage with each needle might have been comparable with its rate at Portsmouth. The accident which prevented the execution of this purpose, and rendered the series of observations much less complete than it would otherwise have been, is much complained of both by Capt. Lütke and M. Lenz. In consequence of this accident, it was not until the arrival of the Siniavin at Kamtschatka that the needles could be vibrated at a station to which they were subsequently brought back; and out of 52 stations, there are only 18 which were observed at during an interval in which anything is known by observation of the steadiness of the magnetism of the needles. They were vibrated at three different dates at the harbour of St. Peter and St. Paul, viz. on September 30, 1827, June 6, 1828, and October 9, 1828. Their changes of rate in the intervals were small, but not proportionate. Corrections are computed and applied at all the intermediate stations in the usual manner. M. Lenz has employed the rate of change of each needle, deduced from the first and second times of vibration at St. Peter and St. Paul, to furnish corrections for the stations observed at antecedently to Capt. Lütke's first arrival at Kamtschatka; of these the land stations are Rio de Janeiro, Concepcion, Valparaiso, Sitka and Unalaska. For a single station (Manilla) observed at subsequently to the final departure from Kamtschatka, M. Lenz has used the rate of correction deduced from the second and third times of vibration there.

The times of vibration were derived on all occasions from the mean of 250 consecutive vibrations, commencing with an arc of 30° and ending usually about 10°. M. Lenz has not considered it necessary to apply a correction for the arcs. The value of the correction to a mean temperature was determined for each needle by observations made at St. Petersburg at the conclusion of the voyage. For four of the five needles the correction was as usual additive to the time for temperatures below the standard, and subtractive for those above it; but one needle, rhomboidal in shape, exhibited the anomaly of a decrease of force in the colder temperatures, fully as great as the increase shown by any of the others. The observations appear to have been very carefully made,—were repeated four times,— and include a difference of temperature of 20° Reaumur. A similar anomaly has been noticed, if I remember rightly, by M.

Kupffer, as having occurred in his experience, and I have my-self met with an instance of the same kind. M. Lenz has em-ployed no correction for this needle; and the vibrations of the vertical needle appear also to have been uncorrected for tem-perature.

The harbour of St. Peter and St. Paul is the fundamental station of Capt. Lütke's determinations. The value of the in-tensity there, 1·447 to 1·348 at Paris, is stated by M. Lenz to be taken on the authority of M. Hansteen.

Capt. Lütke used both his dip and intensity needles at sea in favourable weather, placing the instruments on a board sus-pended in gimbals above the companion. His sea observations appear to be viewed by M. Lenz as not entitled to equal weight with those at the land stations, but as valuable additions. Of 51 intensity results, 16 are at land stations, and are entered in the general table; and I subjoin, as in the case of M. de Hum-boldt's, a separate table of the 35 results obtained at sea.

SOUTH.			1827.		NORTH.			1827.	
29 10	313 35	16 Jan.	(a) 0·924*		0 35	232 56	8 May	(b) 1·013
40 55	307 0	25 Jan.	(a) 1·110		2 24	232 08	9 May	(b) 1·012*
49 18	302 48	31 Jan.	(a) 1·268		13 13	227 0	19 May	(b) 1·112*
53 16	301 37	3 Feb.	(b) 1·320*		23 26	218 02	25 May	(b) 1·212*
55 25	298 27	8 Feb.	(b) 1·413		25 21	213 56	30 May	(b) 1·376*
41 00	282 30	1 March	(b) 1·324		40 28	213 35	1 June	(b) 1·456
29 38	278 26	11 April	(c) 1·153*		44 54	214 50	3 June	(b) 1·573*
21 51	268 05	18 April	(c) 1·046*		48 44	216 37	6 June	(b) 1·653
13 09	251 20	27 April	(c) 1·014		52 29	219 08	9 June	(b) 1·662*
9 38	243 25	30 April	(c) 1·141*		45 27	159 02	23 Oct.	(b) 1·303
6 01	240 08	2 May	(b) 1·005*		39 07	159 03	26 Oct.	(b) 1·186
4 20	238 13	3 May	(b) 0·998		32 59	161 49	1 Nov.	(b) 1·113*
2 29	236 26	4 May	(a) 1·000*		18 44	163 55	13 Nov.	(b) 0·989
2 02	236 04	4 May	(b) 0·996*		11 27	161 52	18 Nov.	(b) 0·970
1 15	225 30	5 May	(c) 0·989*		4 17	162 54	23 Dec.	(a) 1·001
1 10	234 31	6 May	(b) 0·995*		3 47	162 59	23 Dec.	(a) 1·010
0 56	233 17	7 May	(c) 0·990*		2 56	162 50	24 Dec.	(a) 1·013
							1828.	
					6 55	158 02	7 Jan.	(a) 0 990

The results with an asterisk are so marked in M. Lenz's memoir to signify observations made under less favourable circumstances than the others. The sixteen which are not so marked are entered in the general table.

(a) designates results obtained by means of the horizontal needles; (b) those by means of the dipping-needle; and (c) results which are a mean of both me-thods.

King, 1826–1830.—These observations were made during

a survey of the coast of South America from Rio de Janeiro to Valparaiso, carried on under the orders of the British Government by Capt. Philip Parker King of the Royal Navy. They were undertaken at the request of M. Hansteen, and with an apparatus for horizontal vibration with which Capt. King was furnished by him. A copy of the observations was transmitted, from time to time, as they were made, to M. Hansteen, who employed the results, computed provisionally, in his map of the intensity, published in the *Annalen der Physik*, vol. xxviii. The observations themselves have not yet been published, having been given by Capt. King to his successor in the survey, Capt. Fitz Roy, to be published when the latter should return to England. On his return, which took place late in 1836, Capt. Fitz Roy placed Capt. King's magnetic observations in my hands (together with his own, of which a separate notice will be given in the sequel,) to calculate and arrange for publication in an account which he is now preparing for the press, of the proceedings of Capt. King and himself during the survey. Meantime I have Capt. Fitz Roy's permission to introduce Capt. King's results into this memoir.

. The needle with which M. Hansteen supplied Capt. King sustained a very considerable loss of magnetism during the four years it was employed by that officer. Its time of vibration increased between March 22, 1826, and January 24, 1831, (on which days it was tried in the garden of the Royal Observatory at Greenwich,) from 734·5 seconds in 1826, to 775·8 seconds in 1831. A change of such magnitude in the magnetic intensity of the instrument employed to measure the variations of the terrestrial intensity, and which ought itself, therefore, to be invariable, would, in ordinary circumstances, have prevented any satisfactory conclusion whatsoever being drawn from the observations. Fortunately, from the nature of the duties in which Capt. King was engaged, he had occasion to return frequently to the same anchorages; and as he was extremely careful to re-examine the needle on every such return, we have the means of knowing by direct observation the amount of the loss it sustained in certain portions of the time of its employment. There are eleven stations at which the force was observed on the east and west coasts of South America, and two in ports of the Atlantic on the outward voyage. By the practice referred-to, of repeating observations at the same station at distant intervals, the South American stations are so linked together and connected, that by adopting a method similar to that used in determining chronometrical differences of longitude, we may compute and assign the intensity at each, in reference to one selected, and regarded in the same light as a first meridian. In justice to these valuable observations, and in consideration of the great change undergone by the needle, it may be desirable briefly to describe the manner in which this has been done.

At Rio de Janeiro, which was the first station observed at in South America, the needle was vibrated in August 1826, September 1827, and December 1828; in the intervals between these dates are comprised the principal part of the observations on the east side of South America. There is no direct observation at Rio subsequently to December 1828, but we are able to supply the time of vibration at a fourth date in the following manner. The intensity at Rio and at Monte Video having been correctly compared by a double comparison in 1827 and 1828, the needle was vibrated at Monte Video on the 1st of June, 1830, immediately before Capt. King's departure for England, and we thus obtain by an easy calculation the time of vibration at Rio corresponding to the same date. The intervals between these four dates include the whole of the South American stations; and we have only to distribute in each interval the loss of magnetism which the observations show to have taken place from one date to the next, in the manner which may appear most suitable. There is no very obvious indication that the loss was other than gradual; and by considering it uniform in each separate interval, the results are found extremely accordant at several other stations at which observations were repeated at distant intervals. The subjoined tables will enable the reader to judge of this for himself. In the first table are shown the times of vibration at Rio, corresponding to the four dates : 1st, the observed times of horizontal vibration reduced to infinitely small arcs and to a temperature of 60°; and 2nd, the corresponding times as a dipping-needle. The value of the correction for temperature has been determined for this needle by observations which I have recently made with it for that purpose, the particulars of which will be given in the more detailed statement in Capt. Fitz Roy's publication. In the three last columns are shown,—the number of days comprised in each interval,—the increase in the time of vibration owing to loss of magnetism in the needle,—and the resulting daily correction on the supposition of the loss in each interval being uniform.

The second table contains the corrected times of horizontal vibration at each of the South American stations at the dates respectively inserted ;—the dips observed by Capt. King ;—the time of vibration as a dipping-needle at *Rio* at the same dates,

derived from the observations in the first table ;—and the resulting intensity at the station relatively to Rio. The contents of
the tables thus far are the results of Capt. King's observations,
unmixed with those of any other observer. We have now
to express his results in terms of the general scale of comparison, and this is done in the final column, by taking the value
of the intensity at Rio at 0·884, which is the mean of four independent determinations by the following observers, viz. :

$$\left.\begin{array}{l}\text{1817 and 1820 Freycinet . . . 0·890}\\ \text{1827 . . . . . Lütke . . . . . 0·886}\\ \text{1830 . . . . . Erman . . . . 0·879}\\ \text{1836 . . . . . Fitz Roy . . . 0·878}\end{array}\right\}0·884$$

I have included in table II. Madeira and Port Praya, at which
Capt. King observed in his outward passage. The dates of
these fall between the observations at Greenwich in March,
1826, (corrected time = 734·0 and dip 69° 52',) and those at
Rio in August, 1826. Having the intensity at Greenwich
= 1·372 and at Rio = 0·884, we have the time of vibration
as a dipping-needle at Rio at the respective dates as follows :

$$\begin{array}{l}\text{March, 1826 . . . . . . 536·2}\\ \text{August, 1826 . . . . . 537·0}\end{array}$$

It appears, therefore, that only a very slight change took
place in the magnetism of the needle during the outward voyage, and we may take 536·6 as the time of vibration at Rio,
corresponding to the dates of the observations at Madeira and
Port Praya. I have assumed the dip and force at Greenwich
to be the same as at London. The dip at Madeira was not observed by Capt. King, but has been supplied from my own observations in 1822, which were made in the same locality,
namely, the Consul's garden in Funchal, where Capt. King's
needle was vibrated. I have deducted 12' from my determination of the dip at Madeira for the probable change between
1822 and 1826.

TABLE I.

Rio de Janeiro, Dip 14° 90'.	Time of Vibration.		Interval.	Loss.	Per diem.
	Horizontal.	As a dipping-needle.			
	s.	s.	Days.	s.	
August 29, 1826...	545·2	537·0			
			382	6·5	·017
September 15, 1827	551·8	543·5			
			462	9·2	·020
December 21, 1828	561·1	552·7			
			527	2·7	·005
June 1, 1830	563·8	555·4			

TABLE II.

Station.	Date.	Time of horizontal vibration.	Observed dip.	Time as a dipping needle at Rio.	Intensity, Rio =1·000.	Intensity, Rio =0·964.
		s.	° '	s.		
Madeira............	1826, May 31...........	627·79	68 6·0 N.	536·6	1·556	1·377
Port Praya ....,..	1826, June 22 & 24......	557·08	45 44·7 N.	536·6	1·330	1·177
St. Catherine......	1827, Nov. 3	553·58	22 12·4 S.	544·5	1·045	0·920
Gorriti	1826, Oct. 29 & Nov. 6	549·44	35 05·9 S.	538·1	1·172 }	1·041
—................	1829, Jan. 10	562·78	....	552·8	1·179	
Monte Video......	1827, Dec. 18...........	553·87	36 28·4 S.	545·4	1·206	
—................	1828, Oct. 8	560·95	....	551·3	1·201 }	1·065
—................	1830, June 1	564·89	....	555·4	1·202	
Sea-bear Bay ...	1829, March 20	576·37	53 13·5 S.	553·1	1·538	1·361
St. Martin's Cove	1827, Jan. 15 & 22.....	584·29	59 43·8 S.	539·4	1·691 }	1·498
—................	1827, March 27	585·08	....	540·6	1·694	
Port Famine......	1828, Jan. 28	589·36	59 52·6 S.	546·2	1·712	
—................	1828, May 8	596·54	....	548·2	1·683 }	1·505
—................	1828, June 18 & July 20	595·81	....	549·3	1·694	
—................	1830, April 26	598·97	....	555·1	1·712	
Chiloe	1829, Sept. 1 & Dec. 15	565·23	49 52·6 S.	554·2	1·402	1·321
Juan Fernandez .	1830, Feb. 19	551·83	44 49·8 S.	554·8	1·425	1·262
Talcuhuano	1829, Dec. 28...........	555·59	45 10·0 S.	554·6	1·413 }	1·250
—................	1830, May 12...........	557·18	....	555·3	1·412	
Valparaiso........	1829, Aug. 4	548·59	40 10·7 S.	553·9	1·334 }	1·176
—................	1830, Jan. 11 & Feb. 1 .	551·6	....	554·6	1·324	

Sabine, 1827.—These observations were made for the purpose of determining the ratio of the intensity in Paris and London, in order to connect and unite in one system, the results of the different observers who had made Paris and London respectively the base stations of their series.

All values of the intensity hitherto determined are *relative* values; that is to say, each observer has taken some one station as the fundamental one of his series, and has expressed the values of the intensity at all his other stations, comparatively with the force at his fundamental station. Unless, therefore, two such series have one station common to both, or unless the force at their respective fundamental stations has been otherwise compared, they do not form parts of one system, and the results of the one series cannot be brought into connexion with those of the other.

The continental observers in general have taken Paris, either mediately or immediately, as their fundamental station; and the English observers have as generally taken London; the present observations were designed, therefore, as a link to connect their respective series into one system.

c 2

Six horizontal needles were employed for this purpose, and a number of observations were made with them at different dates at both places; the details are published in the *Phil. Trans.* for 1827. From these it appears that, if the horizontal intensity in London be designated as unity, the several needles gave its value in Paris as follows, viz.:

Needle IV. = 1·0732		Needle XI. = 1·0723	
,, VIII. = 1·0675		,, A. = 1·0709	
,, X. = 1·0726		,, B. = 1·0717	

Mean 1·0714.

The observations were corrected for a small excess of temperature in the experiments at Paris over those in London, being, I believe, the first time in which a correction for difference of temperature was introduced into any published results of the variations of intensity at different stations. The places of observation were the magnetic cabinet of M. Arago at Paris, and the garden of the Horticultural Society at Chiswick, near London.

In order to deduce the relative values of the total intensity from their observed horizontal components, we require the dip at the two stations as accurately as it can be inferred from nearly cotemporaneous observations. In August, 1828, the dip in the garden at Chiswick was observed by Mr. Douglas and myself, 69° 46·9. *Phil. Trans.*, 1829. In a paper of M. Hansteen's, in the *Annalen der Physik*, vol. xxi. p. 414, we find recorded the following observations at Paris, a part of which fall on either side of the London observation of August, 1828, viz.:

1825 Arago	68°	00
1826 Humboldt and Mathieu . .	67	56·5
1827 Humboldt and Mathieu . .	67	58·0
1830 Arago	67	41·3

The months in which the observations were made are not named by M. Hansteen, but M. de Humboldt in a paper in the xvth vol. of the *Ann. der Physik* mentions that those of 1825 and 1826 were made in August and September, and I have taken those of 1827 and 1828 as corresponding to the same months. Allowing then an annual decrease of dip of 2'·8 (*Ann. der Physik*, vol. xxi. p. 419) we obtain the dip in Paris in August, 1828, as follows:

	°	'	
1825 Arago	67	51·6	
1826 Humboldt and Mathieu	67	50·9	67° 51·15
1827 Humboldt and Mathieu	67	55·2	
1830 Arago	67	46·9	

I have therefore taken 67° 51'·2 as the most satisfactory cotemporaneous result that I can obtain for Paris, all the observations being made in M. Arago's magnetic cabinet. It appears therefore, that about the period in question, the dip in London exceeded that in Paris by 115'·7; preserving this difference in the dips at the two stations when reduced to the period of the horizontal observations in 1827, and combining them with the observed horizontal intensities, we obtain 1·018 as the value of the total force in London to unity in Paris.

Such being the case, if any other number than unity be taken for the measure of the force in Paris, the corresponding value in London will be the product of that number multiplied by 1·018. By the observations of M. de Humboldt already described, the intensity at Paris to that of a place in Peru, where the needle had no dip, was found to be as 1·3482 to 1·000. As at that period it was supposed that an equal intensity, being the minimum on the surface of the globe, prevailed at all places where the needle had no dip, the station at which M. de Humboldt had observed in Peru appeared the proper unity of the system of intensities. Subsequent experience, however, has shown that the intensity lines follow a very different course from the dip lines; and in retaining the expression of unity for the force observed by M. de Humboldt in Peru, we are necessitated to employ terms less than unity to express the force in many other of the inter-tropical parts of the globe, and even in one quarter beyond the tropic. The scale is therefore purely arbitrary; but it is in general use, and will probably continue to be employed till experiments (perhaps those of M. Gauss) shall have determined an absolute value for the magnetic intensity at some one station; when all the relative intensities may be converted into the corresponding absolute intensities. Such is the origin of the number 1·3482 employed by observers generally as expressing the force at Paris. In assuming a constant expression for the force at any station on the globe for any considerable number of years, we are of course subject to error resulting from the secular change in the intensity; of the amount of which we have as yet no definite knowledge.

The force in London relatively to the above value of the force at Paris is $1·3482 \times 1·018 = 1·372$.

In the spring of 1828 two of the needles used in this comparison were interchanged between M. Hansteen and myself, for the purpose of determining in a similar manner the ratio of the horizontal intensity at London and Christiania. The observations are detailed in the Journal of the Royal Institu-

tion for 1830, p. 29. They gave the following results for the
horizontal intensity at Christiania to unity in London:

Needle IV.	Comparison in March . .	0·9124
	Comparison in May . . .	0·9157
„ VIII.	Comparison in March . .	0·9157
	Comparison in May . . .	0·9160

Mean . . . 0·9147

We have seen that the observations in Paris and London
gave 1·0714 for the horizontal intensity at Paris, also to unity
in London; consequently Christiania to Paris is as 0·9147 to
1·0714, or as 0·8537 to 1. In the spring of 1828 M. Hansteen
observed the dip at Christiania 72° 16'·2; at Paris at the same
time, or about four months before August 1828, we may consider it to have been 67° 52'·5. The total intensity at Christiania derived from this comparison is therefore 1·423. The
result of a direct comparison between Paris and Christiania
made by M. Hansteen in 1825 is 1·419.

All the values of the intensity inserted in this memoir were
originally observed in reference to one of these three stations,
Paris, Christiania, or London, mediately or immediately. They
have been united by means of the comparisons above noticed,
viz., those of Paris and London, and of Paris and Christiania;
and they now form one connected series.

Keilhau, 1827.—These observations were made in a voyage
to Finmarken and Spitzbergen, in which M. Keilhau was furnished with an horizontal apparatus of M. Hansteen's, and a
5-inch dip circle and two needles made by Dollond. The
observations were communicated to M. Hansteen, and the results were published by him in the xivth vol. of the *Annalen
der Physik*, from whence I have taken them.

There may be remarked in these results greater differences
of intensity between stations near to each other than are
usually met with. From the geological character of the countries, it is probable that a portion of these may be due to local
circumstances; but it is also probable that a considerable portion of them may be attributed to the inadequacy of the dipping-needle with which M. Keilhau was furnished, to give results sufficiently exact for the computation of intensities, in a
part of the globe where a small error in the dip will occasion
a very considerable one in the deduced intensity. His two
dipping-needles frequently gave results at the same station
differing from twenty to thirty minutes from each other.

There are 20 stations determined by M. Keilhau in Norway,

Finmarken, and Spitzbergen, all which are inserted in the general table.

Hansteen and Due, Erman, 1828-1830.—In 1819 M. Hansteen published his celebrated work on the magnetism of the earth, in which he brought into one view a larger body of observations of the *dip* and *variation* than had been brought together by any previous philosopher; and by subjecting them to a close examination, drew this remarkable inference in regard to the *intensity*; namely, that a centre, or pole as it might be termed, of magnetic intensity must exist in the north of Siberia, less powerful, but otherwise similar to the one in the north of America; and that the lines of equal intensity would be found to arrange themselves around the Siberian centre in the same way as around the centre of greater force in America. At the time M. Hansteen drew this inference not a single observation of the intensity had been made nearer to Siberia than Berlin on the one side and Mexico on the other.

M. Hansteen's work, much more read on the Continent than in England, produced a very general desire that an inference so remarkable, and so important if confirmed, should be submitted to the test of experiment. This, however, exceeded individual means to accomplish; it was one of those undertakings in science for which national aid is required. To the honour of Norway, the funds for this undertaking were furnished by a unanimous vote of the Norwegian Storthing or Parliament. In 1828 M. Hansteen, accompanied by Lieut. Due, proceeded at his country's expense, and with every facility which could be afforded him by the Russian Government, on a journey expressly for magnetic observations through the Russian dominions in the north of Europe and Asia. They were provided with a dip circle and two needles of Gambey's, and with M. Hansteen's apparatus for horizontal vibrations. At St. Petersburg they were joined by M. Erman of Berlin, proceeding on a similar mission to the same countries, and similarly furnished with magnetic instruments. The three gentlemen travelled together to Siberia, MM. Hansteen and Due on the one part, and M. Erman on the other, making the same observations everywhere, but independently of each other. They wintered at Irkutsk; and the following year MM. Hansteen and Due returned to St. Petersburg by land route, and M. Erman proceeded by Ochozk to Kamtschatka, where he embarked for Europe. The maps attached to this memoir mark by the observations entered on them their various journeys, separately and together, in northern Asia. Suffice it here to say, that they traversed the whole of the north of Europe and of Asia longitudinally, and descended the rivers Oby and Jenesei to the polar circle, with a view of determining the latitude and longitude of the Siberian pole or centre of magnetic intensity; and that its general phænomena were found to correspond in a very remarkable degree with M. Hansteen's anticipations, its locality being removed but a few degrees (about 6°) to the eastward of the position he had previously assigned to it.

Soon after M. Hansteen's return, he published a general map of the magnetic intensity, in the xxviiith vol. of the *Annalen der Physik*. I am not aware that he has as yet published any detailed statement of the results of his journey. The stations inserted in the table in this memoir are from a MS. copy of his and Lieut. Due's observations, which, with the liberality that has hitherto characterised the labours of those engaged in this interesting inquiry, and which I trust may long continue to do so, he sent me from Irkutsk in 1829, with permission to make " every use of it that I might think proper, especially when it can encourage to new undertakings, and accordingly forward the science."

M. Hansteen's determinations of intensity have a very great advantage in the perfect invariability of the needle he employed. For sixteen years in which it was in constant use no sensible alteration took place in its magnetism. This is an advantage which only those can duly appreciate who have been much engaged in making or in computing observations of this nature. The correction for temperature also, which he determined experimentally in the usual manner, has received the fullest practical confirmation, by the exact agreement, when corrected by it, of observations at the same place in temperatures differing nearly 90° of Fahrenheit.

M. Erman's intensity observations are not yet published; they are to form a part of the second volume of the scientific portion of his journey, the first volume of which was published at Berlin in 1835. He has, however, communicated their results, provisionally computed, with corrections applied for temperature and arc, in the xviith vol. of the *Annalen der Physik*, from whence I have extracted them.

The number of stations entered in the table are, 80 observed by MM. Hansteen and Due, and 98 by M. Erman. These are all in the north of Europe and Asia, and 46 are common to M. Erman and MM. Hansteen and Due. There are besides four land determinations of M. Erman's on his homeward voyage, viz., Sitka, St. Francisco in California, Otaheite, and Rio de Janeiro. He made also a very extensive series of intensity observations on board ship in his passage from Kamtschatka to Europe. Of these he has not yet communicated the numerical results. He observed the vibrations of a dipping-needle placed on an apparatus contrived to guard against the ship's motion, which is understood to have been very successful*.

Kupffer, 1829.—These observations were made in a scientific journey to the Caucasus, undertaken by the order of the Emperor of Russia. M. Kupffer was furnished with two horizontal needles, one of which he received from M. Hansteen, and the other from myself through M. de Humboldt. He employed them, between May and August, 1829, at St. Petersburg, Moscow, Stavropol, two stations in the Caucasus, Taganrog, and Nicolaieff; and on his return to St. Petersburg, presented to the Imperial Academy of Sciences a report on the general results of his journey, in which the times of vibration of the needles are specified, together with the temperatures and the observed dips; but the conclusions, in regard to the relative intensity at the different stations, were deferred, until the corrections for temperature for the two needles could be experimentally investigated. I am indebted to M. Kupffer for a printed copy of this report, and I have

---

* Since this report passed from my hands into those of the Assistant-general Secretary, I have been favoured by M. Erman with a complete copy of his observations, including those made at sea. On hearing from M. de Humboldt that I was engaged in drawing up this report, M. Erman, with great liberality and most obligingly, sent me a copy in manuscript of the whole of his results provisionally computed. I have thus been enabled to add five or six stations between Ochozk and the harbour of St. Peter and St. Paul with which I was previously unacquainted, and 167 observations made on his voyage from Kamtschatka to Europe. I consider these last observations particularly valuable, in the evidence they afford, that determinations of the intensity can be made at sea with an accuracy but little inferior to those on land. With the exception of a few in the very early part of the voyage, which appear from some cause to give somewhat lower intensities than accord with M. Erman's own observations at Sitka and St. Francisco, the results, both in the Pacific and Atlantic, whenever they approach the land stations of other observers, present a most satisfactory accordance.

The complete series of M. Erman's magnetic determinations is the most extensive contribution yet made to the experimental department of magnetical science ; nor can we rate its value too highly, since it furnishes us with consecutive determinations of dip, variation, and intensity, by the same highly qualified observer, and with the same excellent instruments, extending through all the meridians of the globe, and from the Arctic circle in Siberia to nearly 60° of south latitude, the whole of this distance being traversed in the course of two years, and the track completely marked by the frequency of the observations.

seen no later publication containing his own conclusions from
his observations. The results entered in the table are consequently computed by myself from the report above noticed, and are uncorrected for temperature, which is of the
less importance as the differences of temperature were not
considerable. It is not stated in the report that the needles
were re-examined at St. Petersburg at the close of the series;
but as the two give results very nearly accordant, it is probable they underwent little or no loss. At one of the stations in the Caucasus no dip was observed ; consequently no
total intensity can be computed. Some error has obviously
taken place in regard to the observations at Moscow ; the
times of vibration of both needles as given in the report would
correspond with a very much higher intensity there than at St.
Petersburg, which we know from the concordant observations
of MM. Erman and Hansteen is contrary to fact. M. Hansteen, who received the observations direct from M. Kupffer
at St. Petersburg, has omitted the Moscow results in his notice
of this series. I have therefore done the same, supposing that
there is some satisfactory reason for the omission with which
I am unacquainted. At Stavropol and Taganrog the dips
employed in the reduction were observed with an inferior instrument, the principal dipping-needle having met with an accident.

Quetelet, 1829–1830.—In 1829 M. Quetelet, Director of the
Royal Observatory at Brussels, made observations on the horizontal intensity at several stations in Germany and the Nether-
lands, with an apparatus similar to M. Hansteen's and two
needles ; and in the following year in France, Switzerland, and
Italy with the same apparatus and four needles. The observations of 1829 are contained in a memoir printed in the 6th vol.
of the *Memoires de l'Academie Royale de Bruxelles* ; those of
1830 in the *Annalen der Physik*, vol. xxi. Unfortunately, the
greater part of the observations of horizontal intensity are unaccompanied by observed dips, and the stations are comparatively few at which M. Quetelet either observed the dip himself
or has selected dips observed by others, so as to be available
for our present purpose. There are ten such stations entered
in the general table. Having vibrated his needles in Paris in
1830, the values of the intensity are deduced by direct comparison. He has corrected the observations for temperature,
employing for their reduction the coefficient determined by M.
Hansteen for his own needle.

Douglas, 1829–1834. —These observations were made by Mr. David Douglas during a journey to the N.W. coast of America, undertaken for botanical and geographical objects. The circumstances of his much-regretted death at Owhyhee in the spring of 1834, whilst waiting for a vessel to convey him home to England, are too well known to need repetition here. Having been supplied with instruments for a part of the scientific purposes of his journey by the Secretary of State for the Colonies, his papers on such subjects were sent by the British Consul at the Sandwich Islands to the Colonial Office, and on their arrival in England were placed in my hands to examine and report upon. The books containing the magnetical observations showed, by the completeness of the record, the attention and care bestowed on every circumstance which could conduce to accuracy. A full report on these, and on his other scientific papers, has been presented to Lord Glenelg, the present Secretary of State for the Colonies, but is yet unpublished. I have therefore permitted myself to enter into a more circumstantial account of these observations in this place than I have done in regard to other observers, whose works can be immediately consulted.

Mr. Douglas was furnished with a dip circle of $11\frac{1}{2}$ inches in diameter, made by Dollond, with a needle on Mayer's principle ; and for the intensity, with four of the same horizontal needles which I had used in 1822–1823, viz., Nos. 3, 4, 5, 6. The time of vibration of these needles was observed by Mr. Douglas in London, in 1828 and 1829, previously to his leaving England. In May, 1830, they were vibrated at Oahu, one of the Sandwich Islands; and between September, 1830, and February, 1831, at four stations in North America, where the dip was also observed, viz., Fort Vancouver, Cape Disappointment, Monterey, and St. Francisco ; and at several other stations, where the dip was not observed. In February, 1831, he sent Nos. 3 and 4 to England, to have the permanency of their magnetism examined ; retaining Nos. 5 and 6 with him for further observations. Nos. 3 and 4, from accidental circumstances, did not reach me till 1836 in Ireland, and being examined in Limerick and Dublin (both which stations had been carefully compared with London), No. 3 was found to have slightly gained, and No. 4 slightly lost magnetism, on a comparison with their rates in 1828 and 1829. When not employed in actual observation, these needles were kept together in the same case, with their opposite poles connected, as were Nos. 5 and 6 in another and a separate case. I have had occasion to remark elsewhere, that, when needles differing considerably in their rates of vibration are so kept together, it does not unfrequently happen that the weaker needle acquires magnetism, and the stronger loses it; and such appears to have been the case in this instance. It was not until 1829 that Nos. 3 and 4 were put together, having been previously paired in a similar manner with other needles, whose magnetic strength in both cases very nearly coincided with their own. It is probable, therefore, that the one began to lose and the other to gain from that time forth; and that the whole gain or loss took place in the earlier portion, rather than equably throughout the interval from 1829 to 1836.

When needles are so kept together in pairs, the two should be employed on every occasion, and their combined result should be regarded as one determination. Mr. Douglas never employed them singly. If in such cases the gain of the one needle were exactly proportioned to the loss of the other, the results of the two needles taken separately would differ, but combined would furnish a mutual compensation. In the present case the gain and loss, though not identical, were so nearly equal, that by taking a mean between the London rates of each needle in 1829 and 1836, and combining at London and at the other stations the results of the two needles into one determination, we obtain the values of the intensity as they would have been given by a single needle whose magnetism had undergone little or no change.

The intensities thus calculated by needles 3 and 4, for the Sandwich Islands and the stations in North America, are almost identical with those computed from Nos. 5 and 6, taken jointly in the same manner, using the London rates which they had before they left England. These needles have been sought for in vain amongst Mr. Douglas's effects sent to England; their steadiness, therefore, can only be judged of from a comparison of their results with those of Nos. 3 and 4.

The special objects of Mr. Douglas's mission leading him in excursions on foot into the interior of the country, in California, and on the rivers tributary to the Columbia, the use of the horizontal needles was the only service he could there render to magnetism, as the dip circle was not sufficiently portable to be taken with him. There are 18 stations at which he used the horizontal needles alone, between $34\frac{1}{2}°$ and $54\frac{1}{2}°$ N. lat., and all nearly on the same meridian, viz., between $119°$ and $124°$ W. from Greenwich. The only absolute deduction in these cases is that of the horizontal intensity. In deducing the total force from its horizontal component, the dip employed must necessarily be computed from the dips observed at other stations. Determinations of intensity in that part of the globe are as yet so rare, that such observations are too valuable to be omitted in this memoir; I have accordingly entered them in the general table, as well as in a separate table here, and have annexed to the latter a brief notice of the manner in which they have been computed.

The last observations recorded in Mr. Douglas's books are those which he made on the dip at Byron's Bay, and on the force, with needles 5 and 6, at Byron's Bay and in the crater of the volcano Kiraueah, soon after his arrival at Owhyhee in 1834. I have searched in vain, amongst the few loose papers which were sent home, for the rough notes of observations of very great interest, of which he speaks in his private letters, but which are not entered in his books. I mean those of the dip, variation, and intensity at the summit of Mowna Kaah, nearly 14,000 feet above the sea, and at other elevations on the island exceeding 10,000 feet. He mentions, as a general inference from these observations, that he found little or no difference between the three phænomena observed at those heights and near the sea. Those in the crater of Kiraueah, about 4000 feet above the sea (which are the only ones preserved), indicate a decidedly less intensity (1·059 to 1·098) than on the sea side at Byron's Bay, a few miles distant: but Kiraueah is a recent volcano, and no conclusion, as to the simple effect of elevation on the magnetic intensity, can of course be drawn.

In the first subjoined table are inserted the intensities determined at the stations where both the dip and horizontal intensity were observed. The second table contains those stations where the horizontal component only was observed, and the dips are supplied in the third table according to the explanation annexed to it.

TABLE I.

Station.	Date.	Lat.	Long. west from Greenwich	Dip observed.	Intensity. London=1·372.		
					Nos. 3 & 4.	Nos. 5 & 6.	Mean.
Fort Vancouver...	Nov., 1830...	45 37	122 36	69 39·7	1·684	1·691	1·688
Cape Disappointment	Sept. Dec., 1830.........	46 16	123 56	69 30·3	1·668	1·679	1·674
Point George ...	....	46 11	123 40	69 16·8	....	....	....
St. Francisco......	Feb., 1831 ...	37 48	122 25	62 58·0	1·597	1·597	1·597
Monterey	Jan., 1831 ...	36 35	122 0	62 07·5	1·584	1·596	1·590
Owhyhee	Feb., 1834...	19 43	156 10	37 58·0	....	1·098	1·098

TABLE II.

Monterey = 1·000.		Fort Vancouver = 1·000.	
Place.	Horiz. Int.	Place.	Horiz. Int.
Stuart's Lake	0·5616	Mouth of the Wul-lawullah	0·9790
Frazer's Lake	0·5719		
Fort Alexandria ...	0·6015	Rapids of the Co-lumbia	1·0000
Thompson's River	0·6415		
Oakanagan	0·7165	South branch of the Multnomah	1·0163
San F. Solano	0·9721		
San José	0·9859	Sandiam River......	1·0463
La Soledad	1·0056		
San Antonio.........	1·0080		
San Miguel	1·0101	London = 1·000.	
San Obispo	1·0222		
Santa Barbara	1·0413	Oahu..................	1·758
Santa Ynez	1·0335	Kiraueah	1·762
La Purissima	1·0282		

TABLE III.

The latitudes in this table, and the longitudes of the stations on the River Columbia and its tributaries are from Mr. Douglas's observations. The longitudes are chronometrical, from Fort Vancouver as a first meridian. The longitude of Fort Vancouver is computed from 1200 lunar distances observed by him. A few

Notice of the manner in which the results in the above table have been computed.—There are five stations in North America at which Mr. Douglas observed the dip. The number of separate observations is 21 distributed as follows:

Cape Disappointment	3
Point George	2
Fort Vancouver	6
St. Francisco	3
Monterey	7

To compute from these the dip at the eighteen stations where it was not observed, we require the direction of the isoclinal lines, and the rate at which the dip increases in the perpendicular to them.

The relative position of the five stations, being nearly on the same geographical meridian, is unfavourable for determining the direction of the lines; but, on the contrary, extremely favourable for a deduction of the rate at which the dip increases in the perpendicular to them; and as the horizontal stations are all nearly under the same meridian also, the rate of increase is the element of calculation, which it is most important to obtain correctly.

To compute, therefore, the rate of increase from the observations themselves, we may take the direction of the lines from a general map, as a small uncertainty in this respect has little influence on the result. In M. Hansteen's map of the lines of dip in 1780 we find their direction in that part of the globe to be from N. 74° W. to S. 74° E.* If we express by r the rate of increase corresponding to a geographical mile, and make $\delta =$ the dip at a central geographical position, say 45° N. lat., and 124° W. long., and δ_1, δ_2, δ_3, &c., the observed dip at the five stations, we shall have

$$\delta_1 = \delta + (a_1 \cos 74° - b_1 \sin 74°)\, r$$
$$\delta_2 = \delta + (a_2 \cos 74° - b_2 \sin 74°)\, r, \&c.,$$

the coefficient a being the difference of longitude between the central station and that at which the dip was observed, ex-

---

* When I wrote the above I had not seen M. Erman's more recent magnetic map from his own observations in 1828, 1829 and 1830, in which are delineated the dip lines of 60°, 65°, and 70°, which pass through the district in which Mr. Douglas's observations were made. Their direction in the meridian of 124° W. measured on M. Erman's map is, as nearly as the measurement can be made, from N. 74¼° W. to S. 74¼° E. I add this note to explain the reason why the direction in the text was not taken at once from the more modern and cotemporaneous map, and to express the satisfaction I feel in this confirmation of the element I had ventured to introduce for the calculation of Mr. Douglas's results,—the only element in the calculation which was not furnished by his own observations.

pressed in geographical miles, and b the difference of latitude also in geographical miles.

If we combine the five equations so formed for the five dip stations by the method of least squares, giving each equation a weight proportioned to the number of observations which it represents, we obtain by the usual process of summing and elimination

$$\delta = 68° 42'\,; \quad r = -0.013608,$$

the latter being equivalent to 73.5 geographical miles to one degree of dip. With these we may compute the dip for each of the horizontal stations; and having the values of the horizontal component we may deduce the total intensity. The dips and intensities for the North American stations in Table 3 are thus computed.

Mr. Douglas mentions that the dip he observed in the crater of Kiraueah was 2' greater than at Byron's Bay; I have therefore entered it in Table 3 as 38° 00'. The dip at Oahu is from Capt. de Freycinet's observations at the adjacent island of Mowi, and must be regarded as uncertain for Oahu to some minutes; but in so low a magnetic latitude an error of that amount would have very little influence on the calculation of the intensity. The horizontal intensity at Oahu was very well determined, the four needles being employed, a few months only after their vibration in London.

Fitz Roy, 1831–1836.—We come next to a series which must rank amongst the most important contributions to magnetical science, and which we owe to Capt. Fitz Roy, R.N., and the officers of H.M. ship Beagle, employed in the years above-mentioned in the survey of the coasts of South America, and in a voyage of circumnavigation performed chiefly in the southern hemisphere, having for its primary object the determination of differences of longitude by a number of chronometers.

Capt. Fitz Roy had the precaution to furnish himself with a dipping needle of Gambey, whose instruments of this kind, though not always without fault, are universally acknowledged to be the best that are made, and superior to those of our own artists in modern times. For the intensity he received from Capt. King the horizontal needle with which that officer had been supplied by M. Hansteen. This needle, which in Capt. King's voyage had lost from time to time considerable portions of its magnetism, appears to have very nearly attained a permanent magnetic state when Capt. Fitz Roy received it. By observations at Plymouth in 1831 and 1836, and at Port Praya in 1832 and 1836, its time of vibration is shown to have varied to a very

inconsiderable amount, admitting of safe and easy interpola-
tion.

Capt. Fitz Roy's observations are not yet published. On his
return to England he paid me the compliment of placing them
in my hands to calculate and arrange for publication in the
appendix of an account of his voyage, which he is preparing.
Meanwhile he has permitted me to insert the intensity results in
the general table of this memoir. They are corrected for temperature and for arc. They include 27 stations, of which 24
in the southern hemisphere, distributed throughout its longitudes, throw very considerable light on the system of the intensity in those regions. This extensive series is, I trust, but the
precursor of what British naval officers will accomplish for magnetism in the southern hemisphere.

Rudberg, 1832.—These observations were made with a dipping-needle and two horizontal needles of Gambey's, at five
stations on the continent of Europe, of which Paris was one.
A full account of them is published in the xxviith vol. of the
Annalen der Physik. They appear to have been made with
great care, and the results are corrected for temperature.

Lloyd and Sabine, 1835–1836.—These observations were
made in compliance with a wish expressed by the British Association that some of its members would undertake a survey of
the dip and intensity in the British Islands. Accordingly the
intensity was determined at 30 stations in Ireland by Mr. Lloyd
and myself, in 1835, and by myself at 25 stations in Scotland,
in 1836. The volumes of the Reports of the British Association
for those years contain a full account of these observations, as
well as of the mode in which the determinations at the several
stations are all made to concur in assigning the intensity at
one central position in each country as their general result.
It appears unnecessary, therefore, to reprint them in this
volume, and it is only the intensities at the central position,
thus calculated, which are entered in the general table.

Ross, 1836.—These observations were made in a voyage to
Davis's Straits, undertaken by Capt. James Ross, R.N., in
the winter of 1836, to relieve the crews of several whalers
which had been detained in the ice. Those of the intensity
were made with two horizontal needles in an apparatus similar
to M. Hansteen's. The magnetism of one remained quite
steady during the voyage; the other sustained a slight loss,
which it is evident by inspection took place between Orkney

and Greenland, and has been allowed for accordingly; Orkney being compared with the first London rate, Greenland and Labrador with the second. The needles then give every-where very nearly identical results.

The dip circle which Capt. Ross employed was of 4 inches diameter. The needle appears to have given very consistent results always at the same station; for example, of six obser-vations at Westbourn-green near London in 1836, the ex-tremes are 69° 28' and 69° 35'·6, the poles being changed in every observation; the mean of the six, however, as well as each of the separate results, is a few minutes higher than the dip at that spot is known to have been at that time. Taking into account Capt. Ross's experience in observations of this kind, and that the observations were made on four different days, it is most probable that there was some instrumental cause for this needle giving constantly at this station a higher dip than the truth. Being ignorant, however, what that cause may have been, I have not ventured to apply a correction to the dips with this needle either there or elsewhere, but have em-ployed them just as they were observed at each of the stations.

In countries where the dip is so great as in the vicinity of Davis's Straits, the horizontal intensities may be very correctly determined, and yet from slight errors in the dip, the resulting total intensity may present anomalies unusual elsewhere. We have an instance of this in Capt. Ross's observations in Green-land. There are two stations in Greenland, at no great distance apart, where the difference of the computed intensity is excess-ive; and the fact of there being some anomaly in the observed *dips* which would sufficiently explain the difference, is made quite obvious by the circumstance that the higher dip is at the southernmost station; whereas the dip should increase in going northward on this coast, and with this the horizontal vibrations are in accord. I have therefore omitted both the results in Greenland in the general table.

As these observations have not been published elsewhere, I subjoin a table containing the principal particulars.

Station.	Date.	Lat.	Long.	Time of horiz. vibra.		Dip observed.	Intensity London = 1·372.
				No. 1.	No. 2.		
London......	Aug., 1835	51 31	359 50	439·07	441·46	....	1·372
Stromness...	Feb., 1836	58 58	356 30	480·22	483·34	73 36	1·419
Greenland {	June, 1836	66 57	306 26	648·57	645·30	82 51	1·798
	June, 1836	68 59	306 47	667·29	665·94	82 23	1·590
Labrador ...	Aug., 1836	57 33	298 2	616·11	....	73 36	1·682
London......	Oct., 1836	51 31	359 50	442·19	441·64	69 32·1	1·372

The times of vibration are reduced to a standard temperature.

Estcourt, 1836.—These observations were made during the late survey of the navigation of the River Euphrates, conducted by Colonel Chesney. The magnetic observations were entrusted to Major Estcourt, who was furnished with a good dip circle by Robinson, and an apparatus similar to M. Hansteen's, with eight horizontal needles. Numerous observations were made with these at Port William and Bussora, the manuscripts of which have been sent to me, by the President of the Board of Control, to arrange for publication in the official account of the proceedings of the expedition, preparing under the direction of Colonel Chesney. On the arrival in England of the needles, which only took place very recently, they were also placed in my hands, in order that the necessary comparative observations might be made with them. It had unfortunately happened that the manuscript containing the times of vibration of the needles observed by the officers of the expedition before its departure from England, were on board the Tigris steamer when she was lost in the Euphrates, and no record was preserved. But on receiving the needles, I recognised two of the number as having belonged to Professor Lloyd, of Dublin, and as having been employed by Mr. Lloyd and myself in Ireland. I had consequently· a memorandum of their rates before they were given to the officers of the expedition ; and on vibrating them in Sussex, where I was staying when I received them, I perceived with great satisfaction that these two needles must have preserved their magnetism wholly or very nearly unaltered. They were immediately sent to Professor Lloyd, who kindly vibrated them at the same spot in which they had been used in 1834, and found their magnetism almost identical with what it had been at that period. On trying the six other needles, I found that two gave similar values for the intensity at Port William and Bussora with those of Mr. Lloyd; whence I inferred that those also had undergone no change in their magnetism since the observations on the Euphrates. The determinations at Port William and Bussora inserted in the general table of this report are derived from these four needles. Their times of vibration have been reduced to a standard temperature, the coefficient in the formula having been ascertained for each needle by experiments made since they have been placed in my hands. The full details will be communicated in Colonel Chesney's official publication.

Freycinet, 1817–1821.—I am most happy in being able to add to this collection the valuable observations of Capt. de Freycinet in the voyage of circumnavigation, performed in the

Uranie in 1817–1821. Having heard that I was engaged in drawing up this report for the British Association, Capt. de Freycinet, unsolicited, did me the honour to propose to place his observations, hitherto unpublished, in my hands, to be communicated to the public through this channel. I should certainly fail if I attempted to express my sense of this act of great liberality; happily it needs no comment; and I will only observe, that it adds another instance, but a very strong one, to those already noticed, of the good feeling that has prevailed amongst the persons by whom these inquiries have been carried forward. The world hears more than enough of the jealousies and enmities which too often disfigure the history and embitter the pursuits of science; it is right that the instances to the contrary should not always be passed in silence.

The manuscript of the observations was accompanied by the following remarks from Capt. de Freycinet.

" J'ai mis une grande attention à ce qu'il ne se glissa pas de faute dans la copie; et telle qu'elle est je crois que vous pouvez compter sur son exactitude. L'experience a prouvé que les aiguilles Nos. 7 et 8, dont je me suis servi, ont perdu un peu de leur magnétisme pendant le voyage; il sera facile d'en tenir compte, comme aussi des légères altérations qui auront eu pour cause les variations de température; mais je ne me suis pas livré à ces considérations, pensant qu'il valait mieux que vous vous en occupassiez selon vos vues particulières."

The table in pages 38 and 39 contains the observations, printed from this manuscript without alteration of any kind.

In compliance with the wish expressed by Capt. de Freycinet, I proceeded to calculate the results of these observations in the following manner. The consideration of No. 9 was put aside in the first instance for the reason assigned in the marginal note to the observations at the Isle of France. The times of vibration at Paris before and after the voyage, confirmed by the observations at Rio de Janeiro in 1817 and 1820, show that Nos. 7 and 8 both slightly lost magnetism, and No. 8 rather more than No. 7. It further appears that the extra loss of No. 8 over No. 7 was all sustained in the first fourteen months; as at the Isle of France in June, 1818, they had arrived nearly at an equality in their time of vibration, which they preserved for the whole remainder of the voyage, and exhibited on the return to Paris. In whatever way, therefore, we may proportion the *equal* loss sustained by both needles, the *extra* loss of No. 8 must be placed before the arrival at the Isle of France. When there are no circumstances in the observations themselves indicating otherwise, the usual course is to distribute a loss equally through the interval in which it is known to have occurred. I have therefore pursued this course in regard to the loss sustained by No. 7; and in the case of No. 8 I have allowed a double proportion in each of the first fourteen months. The observations furnish two tests of the propriety of this distribution: the general agreement of the results of the two needles with each other at the different stations is one; the other is the agreement of the force thus calculated at Rio in 1817 and 1821. In both the accordance is satisfactory.

On computing the intensity at the Cape of Good Hope and the Isle of France by No. 9, using for that purpose its time of vibration at Paris in 1817, the results appeared to agree extremely well with those of Nos. 7 and 8. It is hence inferred, that until the accident at the Isle of France, No. 9 had undergone no change of magnetism, and I have therefore brought into the account all the results obtained with it before that occurrence. As the effect of changes of temperature on these particular needles does not appear to have been ascertained experimentally, no corrections are applied on account of temperature; but, as I have before remarked, such corrections are of minor importance in so extensive a series as the present. The table in page 40 exhibits the computed results, and appears to need no other explanation, except that the column entitled "Time of vibration as a dipping-needle at Paris" exhibits the times of vibration corrected for loss of magnetism.

Résumé des Observations d'Intensité Magnétique, faites pendant le Voyage de l'Uranie autour du Monde.

Localité.	Position géographique.		Époque.			Inclinaison moyenne.	Déclinaison moyenne.	Intensité magnet.		Remarques.
	Latitude.	Longitude, comptée de Paris.	Date.	Heure de l'observation d'intensité.	Température moy. centig. de l'air.			No. de l'aiguille observée.	Durée de 100 oscill. Intsn. petites.	
Paris, avant le départ.	48 50 15 N.	0 0 0 O.	1817 Avril 4	3 20 s.	17·95	68 28 28	22 25° 0 N.O.	7	1019·6	*Déclinaison observée en 1816. N.B. No. 7 a été faite
			Avril 28	11 11 m.	8·80			7	1019·9	
			Mars 30	9 4 m.	13·25			8	1009·6	
			Avril 4	4 6 s.	17·90			8	1009·6	
			Avril 28	10 32 m.	8·46			9	1010·6	
			Avril 29	11 43 m.	14·30			9	524·7	
			Avril 29	midi 19 s.	14·60				525·4	
Paris, au retour	id.	id.	1821 Avril 16	11 32 m.	9·5			7	1042·8	
			Avril 16	midi 33 s.	9·4			7	1044·0	
			Avril 16	1 1 s.	9·5			8	1045·3	
St. Croix de Ténériffe	28 27 57 N.	18 35 8 O.	1817 8bre 26	1 30 s.	23·0	57 56 40	21 3 55 N.O.	9	450·7	1 mot ou 1 mr.
			8bre 26	2 0 s.	23·2			9	450·8	
Rio de Janeiro, 1re Relâche	22 55 1 S.	45 38 52 O.	8bre 23	3 30 s.	23·0	14 42 14	2 14 40 N.E.	7	775·9	Temps orageux, tonnerre dans le N.O.
			Xbre 24	8 0 s.	24·1			8	766·7	
			Xbre 24	8 50 m.	26·2			8	768·2	
			Xbre 24	9 30 m.	28·8			9	766·9	
			Xbre 24	11 30 m.	24·3			9	402·6	
			Xbre 28	midi 0	24·3				401·8	

Lieu	Latitude	Longitude	Date	Heure						Obs.
Rio de Janeiro 2ième Relâche	22 55 25 S.	45 38 23 O.	1820 Aout 22	1 53 s.	21·5	14 49 43	3 34 12 N.E.	8	791·2	
			22	2 13 s.	21·5			8	790·6	
			22	3 30 s.	21·5			7	790·5	
Cap de Bonne Espérance	33 55 15 S.	16 3 45 E.	1818 Mars 31	8 30 m.	23·0	50 47 3	26 30·31 N.O.	9	477·9	
			31	11 45 m.	27·0			8	937·0	
Ile de France (Port Louis)	20 9 56 S.	55 8 26 E.	Juin 22	midi 40	29·9	55 6 45	12 46 26 N.O.	9	468·3	
			22	3 29 s.	30·1			9	467·6	
			26	3 14 s.	29·5			7	467·5	
Baie des Chiens-marins (Nlle Hollande)	25 43 21 S.	110 59 13 E.	7bre 30	1 40 s.	29·0	54 52 45	3 38 4 N.O.	8	918·0	
			24	midi 25	29·1			7	913·0	
			24	11 30 m.	20·5			7	809·5	
Ile Timor (Coupang)	10 9 56 S.	121 15 22 E.	8bre 24	1 40 m.	21·8	39 52 3	0 13 38 N.O.	7	799·4	
			19	8 15 m.	21·6			7	799·5	
			19	9 90 m.	29·1			8	809·3	
Ile Rawak (Iles des Papous)	0 1 34 S.	128 35 5 E.	Xbre 30	4 31 s.	29·8	14 26 57	1 29 52 N.E.	8	729·5	
			31	10 36 m.	29·2			8	729·7	
			31	11 26 m.	29·5			7	721·6	
					30·3			7	721·6	
Isles Mariannes (Agagm)	13 27 51 N.	143 37 25 E.	1819 Mai 24	4 13 s.	31·7	19 46 53	4 39 17 N.E.	8	749·9	
			25	8 4 m.	29·1			7	749·2	
			25	8 50 m.	29·1			7	749·1	
Ile Mowi (Sandwich) Rahetra	20 52 7 N.	159 2 3 O.	Aout 22	11 40 m.	29·1	41 39 22	8 49 20 N.E.	8	793·0	
			22	midi 52	29·9			7	793·8	
Port Jackson (Sydney)	33 51 34 S.	148 48 0 E.	Xbre 22	10 38 m.	29·7	62 47 7	9 14 36 N.E.	7	846·4	
			22	11 47 m.	30·6			8	847·4	
Il. Malouines (Baie Française)	51 35 18 S.	60 26 52 O.	1820 Aout 11	1 54 s.	13·2	55 20 7	19 25 41 N.E.	7	832·9	
			11	3 44 s.	13·6			8	832·2	

Après cette expérience No. 9 ayant été posée, par inadvertence, très près d'une de nos grands fusseaux magnétiques, s'est trouvée altérée, et a été hors de service.

Station.	Date.	Needle.	Time of vibration.			Intensity.	
			Horizontal.	As a dipping-needle.		Paris = 1·000.	Paris = 1·348.
				At the station.	At Paris.		
Paris..................	1817	7	1019·75	617·7	617·7	1·000	
——..................	,,	8	1009·93	611·7	611·7	1·000 } 1·000	1·348
——..................	,,	9	525·05	318·0	318·0	1·000	
Teneriffe	,,	9	450·75	319·0	318·0	0·9942	1·340
Rio de Janeiro......	,,	7	775·9	763·1	620·0	0·6602	
——...............	,,	8	767·27	754·6	616·7	0·6679 } 0·658	0·867
——:......	,,	9	402·20	395·6	318·0	0·6465	
Cap de Bonne Esperance	1818	8	937·0	745·1	620·9	0·6945 } 0·697	0·945
———	,,	9	477·9	380·0	318·0	0·7005	
Ile de France	,,	7	912·0	689·7	622·8	0·8155	
——......	,,	8	913·0	690·5	622·3	0·8139 } 0·813	1·096
——......	,,	9	467·8	353·8	318·0	0·8082	
Baie des Chiens-marins	,,	7	800·4	607·1	624·0	1·057 } 1·054	1·421
———	,,	8	802·5	608·6	624·2	1·052	
Ile Timor	,,	7	728·5	667·7	624·2	0·8741 } 0·873	1·177
——.........	,,	8	729·8	668·9	624·5	0·8718	
Ile Rawak	,,	7	721·6	710·1	625·1	0·7750 } 0·774	1·044
——.........	,,	8	722·7	711·1	625·5	0·7736	
Iles Mariannes ...	1819	7	749·15	739·8	626·9	0·7181 } 0·718	0·968
——...	,,	8	749·9	740·6	627·5	0·7180	
Ile Mowi............	,,	7	792·8	685·3	627·9	0·8395 } 0·840	1·133
——............	,,	8	793·0	685·5	628·5	0·8401	
Sydney	,,	7	846·4	572·4	629·5	1·210 } 1·210	1·631
——...............	,,	8	847·4	573·1	630·3	1·210	
Iles Malouines ...	1820	7	832·2	627·6	630·8	1·010 } 1·011	1·363
——...	,,	8	832·2	627·6	631·5	1·012	
Rio de Janeiro ...	,,	7	790·5	777·5	632·2	0·6612 } 0·662	0·892
——...	,,	8	790·9	777·8	633·6	0·6625	
Paris..................	1821	7	1043·4	635·1	635·1	1·000 } 1·000	1·348
——..................	,,	8	1045·3	636·3	636·3	1·000	

It would have given me great satisfaction had I been enabled to have included in this publication the observations made in India by Capt. Jules de Blosseville, in whose untimely death within the Arctic circle, now, I fear, but too certain, science has sustained the loss of an officer who gave full promise, had he lived, of becoming one of the most accomplished navigators of the age. In the last letter which I received from him, dated at Toulon in 1830, he thus expresses himself in regard to his observations of the intensity :—" Toulon ayant été, et pouvant devenir encore le point de départ de plusieurs expeditions scienifiques, il sérait utile, je pense, d'y connoître d'une manière exacte la valeur de l'intensité magnétique, et je me chargerais volontiers pendant le petit sejour que je vais y faire, d'y observer les aiguilles. Ceci me conduit naturellement à vous parler des observations d'intensité que vous m'avez vues commencer à Paris, et que j'ai faites ensuite dans plusieurs lieux de l'Inde. Si elles avaient été plus satisfaisantes, je vous en aurais entretenu dès mon arrivée; mais malheureusement les aiguilles ont perdu pendant le voyage une partie notable de leur magnétisme, et M. Arago a été d'avis de ne point s'occuper de leurs résultats. C'est ainsi que toutes mes peines ont été perdues, quoique j'eusse eu l'attention de rapporter toutes les observations à Pondicherry, qui était le centre de nos operations, espérant par leur repetition dans le même lieu, connoître le decroissement graduel du magnétisme de nos aiguilles. Si je recommence quelque grand voyage, comme je l'espère, je me livrerai avec plaisir à l'étude de l'intensité, et je m'occuperai à l'avance, de faire faire par Gambey l'appareil de plus commode. Je voudrais connoître vos idées sur ce sujet."

Experience has shown in many cases, and particularly in the observations of Capt. King, that it may be possible to obtain very valuable facts from a series of observations, in which the needles have undergone a considerable loss of magnetism in the course of a long voyage ; particularly in cases where attention has been paid to repetition at the same station, for the purpose of a frequent examination of the state of the needles ; and this was practised by Capt. de Blosseville, as well as by Capt. King. Aid may also be sometimes obtained from other observers who may have observed the intensity at some of the stations : and the publication of a series of determinations depending upon Pondicherry would render it an object with persons who might hereafter be engaged in magnetic observations in India, to make Pondicherry one of their stations, and thus supply a link to connect M. de Blosseville's observations with Europe.

In 1833 Mr. Forbes made a very numerous series of excellent determinations of horizontal intensity in different parts of Europe. They were made chiefly with a view to the influence of height on the magnetic intensity, and are discussed in a highly interesting paper in the Edinburgh Transactions for 1836. The dip was observed with a three-inch circle, at a few stations only, and Mr. Forbes has nowhere himself deduced the total intensities. If I am rightly informed, he has since made another tour in the same countries, in which magnetic observations formed a part of his object. We may hope that by a series of dips, corresponding in extent and exactness to his horizontal determinations, he will add greatly to the fulness and accuracy of our knowledge of the course of the magnetic lines in those parts of Europe. The investigation evidently cannot be in better hands. Meantime I have not thought proper to make deductions which he has not made for himself; and the more so, because the stations are very few at which there are both observations of dip and of horizontal intensity, and at some of these the total intensity has already been determined by other observers.

The preceding notices include all the observations of the magnetic intensity with which I am acquainted, in which the instruments, by the steadiness of their magnetism, and their capability of yielding sufficiently precise results, proved worthy of the time and pains bestowed in their employment.

SECTION II.—GENERAL TABLE OF INTENSITIES.

The intensities are arranged in this table according to their values, commencing with those of highest amount in the northern hemisphere, descending progressively to those of least amount, which have their places in the intertropical regions, and again ascending to the highest values in the southern hemisphere. They are classed in zones, the first zone (§ 1) comprehending all the observed intensities in the northern hemisphere between 1·85 and 1·75; the second zone (§ 2), all between 1·75 and 1·65; the third (§ 3), all between 1·65 and 1·55; and so on. In each zone the record in the table commences with the geographical meridian of Greenwich, and passes round the globe in an easterly direction; all the longitudes being counted east from Greenwich, and all latitudes north, unless where it is otherwise distinctly specified.

The geographical position of the several zones is shown in the maps attached to this report by the insertion of the observed intensities themselves in their places in the map. For the more ready guidance and direction of the eye lines are drawn, marking as nearly as can be judged, the middle of each zone. These lines are consequently what are usually denominated isodynamic lines, or lines of equal magnetic intensity at the surface of the earth. They correspond successively to the values of 1·8, 1·7, 1·6, &c., down to 0·8, which is the line of lowest value yet observed. There is, of course, great inequality in the evidence for their precise geographical position in different parts of the globe; sometimes, for the purpose of connection, they have been partially continued where observations are wholly wanting; but in all cases the insertion of the authorities themselves in the map manifests the degree of exactness to which it is yet possible to trace the several portions of each line.

Where the geographical positions are too near each other for convenient insertion in the map, two or more stations are collected into a group in the table, and the mean latitude, longitude, and intensity are placed at the foot of the page. Such groups are in all cases composed of the determinations of the same observer, and the mean determination inserted in the map is characterised by an additional figure, placed beneath, expressive of the number of separate stations thus represented.

In the case of stations visited by two or more observers, their separate determinations have been inserted in the map wherever space has permitted. As this could not always be done in the north of Europe and Asia, the mean of the determinations of the two observers has been given, characterised by the mark +, expressive of the double weight to which such intensities are entitled.

The geographical positions may require correction in a few instances, but pains have been taken to obtain them correctly from the most recent authorities.

DIVISION I. NORTHERN HEMISPHERE.

§ 1. *Intensities from* 1·85 *to* 1·75.

Station.	Lat.	Long.	Observer.	Date.	Intensity.
Viluisk	63 0	120 0	Due..............	1829	1·759
New York	40 43	285 57	Sabine...........	1822	1·803

§ 2. *Intensities from* 1·75 *to* 1·65.

Station.	Lat.	Long.	Observer.	Date.	Intensity.
Turuchansk	65 55	87 33	Hansteen..........	1829	1·667
Sebrinikowo .,..	60 02	90 33	Hansteen..........	1829	1·660
Atschinsk ....'...	56 16	91 00	Hansteen & Due....	1828	1·654
Jenesiek..........	58 27	92 11	Hansteen..........	1829	1·668
Krasnojarsk	56 01	92 57	Erman............	1829	1·652
"	"	"	Hansteen & Due....	1829	1·663
Kansk..........	55 43	96. 53	Erman............	1829	1·670
"	"	"	Hansteen & Due....	1829	1·678
Kamyochatsk......	55 12	98 50	Hansteen & Due....	1828	1·671
N. Udinsk	55 00	99 20	Hansteen & Due ...	1828	1·672

Station.	Lat.	Long.	Observer.	Date.	Intensity.
Kurgan	54 20	100 00	Erman	1829	1·652
Salarinsk	53 30	102 00	Hansteen & Due ..	1828	1·652
Sawaria	53 34	101 53	Erman	1829	1·657
Olonska	52 59	105 04	Erman	1829	1·673
Botowsk..........	55 10	105 22	Erman	1829	1·720
Bojarsk	56 05	105 34	Erman	1829	1·689
Tarakanowa	52 14	106 37	Erman	1829	1·664
Potapowsk	57 17	107 34	Erman	1829	1·711
Kirensk	57 47	108 04	Due	1829	1·704
„	„	„	Erman	1829	1·693
Itschora..........	58 38	109 36	Erman	1829	1·714
Ivanofska	58 38	110 34	Due	1829	1·708
Parchinsk	59 07	111 31	Erman	1829	1·741
Wittinsk.........	59 40	112 00	Due	1829	1·731
Kantinsk	59 53	114 10	Due	1829	1·712
„	„	„	Erman	1829	1·733
Jarbinsk..........	60 28	116 15	Erman	1829	1·702
Beresowsk	59 50	117 56	Erman	1829	1·747
Olekma	60 22	119 33	Due	1829	1·725
„	„	„	Erman	1829	1·707
Sanjacktatsk	60 47	123 46	Erman	1829	1·732
Toen Arinsk	61 37	128 31	Erman	1829	1·689
Yakutsk..........	62 01	129 45	Erman	1829	1·697
Porotowsk	62 01	131 50	Erman	1829	1·721
Lebeghine	62 11	133 42	Erman	1829	1·697
Nokchinsk	61 57	134 57	Erman	1829	1·713
Perewos	61 45	135 40	Erman	1829	1·679
Tchernolies ⎫	61 31	136 23	Erman	1829	1·700
Karnastak ⎭	61 30	137 00	Erman	1829	1·690
Allachjan	61 03	138 45	Erman	1829	1·678
Judomsk..........	60 54	140 35	Erman	1829	1·680
Arki	60 07	142 20	Erman	1829	1·644
Bay of St. Lawrence	65 38	189 14	Lütke............	1828	1·652
At Sea	48 44	216 37	Lütke............	1827	1·653
Sitka	57 03	224 44	Lütke............	1827	1·735
„	„	„	Erman	1829	1·726
Frazer's Lake......	54 03	235 20	Douglas..........	1833	1·724
Stuart's Lake......	54 27	235 40	Douglas..........	1833	1·736
Cape Disappointment	46 16	236 04	Douglas..........	1830	1·674
Fort Alexandria....	52 33	237 31	Douglas..........	1833	1·710

* Mean, 2 stations   61 30   137 00   1·695

Station.	Lat.	Long.	Observer.	Date.	Intensity.
Multnomah River	45 15	237 13	Douglas..........	1830	1·669
Fort Vancouver..	45 37	237 24	Douglas..........	1830	1·688
Sandiam River ..	44 ·35	237 33	Douglas..........	1830	1·683
Columbia Rapids	45 40	238 12	Douglas..........	1830	1·679
Thompson's River..	50 41	239 49	Douglas..........	1833	1·710
Oakanagan........	48 05	240 33	Douglas..........	1833	1·707
Wullawullah River	46 03	241 12	Douglas..........	1830	1·707
Byam Martin's Il. ..	75 10	256 16	Sabine	1819	1·653
Regent's Inlet......	72 45	270 19	Sabine	1819	1·668
Baffin's Bay	76 08	281 39	Sabine	1818	1·659
Baffin's Bay	76 45	284 00	Sabine	1818	1·666
Baffin's Bay	70 35	293 05	Sabine	1818	1·661
Labrador	57 33	298 09	Ross	1836	1·682

§ 3. *Intensities from* 1·65 *to* 1·55.

Station.	Lat.	Long.	Observer.	Date.	Intensity.
Spitzbergen, Fairhaven	79 40	11 40	Sabine	1823	1·562
Spitzbergen, South Cape	76 35	14 00	Keilhau..........	1827	1·558
Katchegatisk	65 09	65 02	Erman	1828	1·568
Beresow..........	63 56	65 04	Erman	1828	1·580
Kunduwaski	63 18	65 06	Erman	1828	1·584
Wandiask	66 16	65 10	Erman	1828	1·608
Kondinsk	62 13	66 36	Erman	1828	1·596
Obdorsk..........	66 31	66 42	Erman	1828	1·580
Jugakow..........	57 32	67 06	Erman	1828	1·546
,, ...	,,	,,	Hansteen & Due ..	1828	1·558
Chutarbitka	57 59	67 31	Erman	1828	1·544
,, ...	,,	,,	Hansteen & Due ..	1828	1·566
Kewaskirche	61 20	68 05	Erman	1828	1·585
Tobolsk	58 12	68 16	Hansteen & Due ..	1828	1·560
,, ...	,,	,,	Erman	1828	1·554
Samarowo	60 45	68 35	Erman	1828	1·584
Uwatsk	59 00	68 46	Erman	1828	1·564
Kolotschikowo	57 27	68 58	Erman	1829	1·564
Sawotinski........	60 23	69 26	Erman	1828	1·573
Tugalowsk........	59 32	69 40	Erman	1828	1·574
Tara	56 54	74 04	Erman	1829	1·575
Pokrowsk	55 38	77 05	Erman	1829	1·617

* Mean, 4 stations   45 17   237 35   1·680

Station.	Lat.	Long.	Observer.	Date.	Intensity.
Muraschiwa	55 50	76 00	Hansteen & Due ..	1828	1·586
Gotoputowa	55 47	77 00	Hansteen & Due ..	1828	1·577
Autoschina........	55 40	78 00	Hansteen & Due ..	1828	1·585
Kainsk	55 40	78 10	Hansteen & Due ..	1828	1·601
Narym	58 50	81 00	Due	1828	1·638
Tschulum	55 06	81 14	Erman	1829	1·578
Kolyvan..........	55 17	82 45	Hansteen & Due ..	1829	1·611
"	"	"	Erman	1829	1·599
Togursk	58 40	83 00	Due	1828	1·644
Barnaul	53 20	83 56	Hansteen	1829	1·605
Tomsk	56 30	85 09	Erman	1829	1·618
"	"	"	Hansteen & Due ..	1829	1·620
Pojelnik	56 18	87 10	Erman	1829	1·627
Kangatovo........	63 27	87 16	Hansteen	1829	1·648
Irkutsk	52 16	104 20	Hansteen & Due ..	1829	1·642
"	"	"	Erman	1829	1·632
Kadilna	52 07	104 51	Hansteen & Due ..	1829	1·649
"	"	"	Erman	1828	1·634
Chogotsk	53 00	105 00	Due	1829	1·645
Tiumeruska	54 09	105 33	Erman	1828	1·648
Selenginsk	51 20	106 15	Hansteen & Due ..	1829	1·642
Troisko Sawsk	50 21	106 22	Hansteen & Due ..	1829	1·642
"	"	"	Erman	1829	1·628
Monachorowa	50 58	106 29	Hansteen & Due ..	1829	1·624
"	"	"	Erman	1829	1·638
Arsentiska	51 17	106 56	Hansteen & Due ...	1829	1·650
"	"	"	Erman	1829	1·636
Werchne Udinsk ..	51 49	107 47	Hansteen and Due...	1829	1·625
" ..	"	"	Erman	1829	1·626
Ochozk	59 21	143 11	Erman	1829	1·615
Sea of Ochozk	58 46	145 52	Erman	1829	1·677
Sea of Ochozk	58 15	152 01	Erman	1829	1·601
Sea of Ochozk	58 13	157 06	Erman	1829	1·595
Tigil River.......	58 01	158 15	Erman	1829	1·577
Maschura	55 04	158 55	Erman	1829	1·551
St. Croix Bay......	65 28	181 28	Lütke............	1828	1·646
Unalaska	53 54	193 30	Lütke............	1827	1·604
St. Francisco	37 48	235 45	Erman	1829	1·585
"	"	"	Douglas	1831	1·597
⟨ San Solano......	38 17	235 36	Douglas	1831	1·610
• ⟨ Monterey	36 35	236 00	Douglas	1831	1·599
⟨ San José	37 32	236 00	Douglas	1831	1·605
⟨ La Soledad	36 24	236 36	Douglas	1831	1·590

• Mean, 4 stations   37 12   236 03   1·600

Station.	Lat.	Long.	Observer.	Date.	Intensity.
San Antonio	36 01	236 42	Douglas	1831	1·584
* San Miguel	35 45	237 16	Douglas	1831	1·583
St. Louis Obispo .	35 16	237 20	Douglas	1831	1·583
La Purissima....	34 40	237 33	Douglas	1831	1·571
† Santa Ynez	34 36	237 49	Douglas	1831	1·579
Santa Barbara ..	34 25	240 00	Douglas	1831	1·604
Melville Island	74 27	248 18	Sabine	1819	1·624
Winter Harbour ..	74 47	249 12	Sabine	1820	1·638
Possession Bay	73 31	282 38	Sabine	1819	1·637
Baffin's Bay	75 51	296 54	Sabine	1818	1·618
Davis's Straits	64 00	298 10	Sabine	1819	1·621
Baffin's Bay	75 05	299 37	Sabine	1818	1·590
Hare Island	70 26	305 08	Sabine	1818	1·622
Davis's Straits	68 22	306 10	Sabine	1818	1·643

§ 4. *Intensities from* 1·55 *to* 1·45.

Station.	Lat.	Long.	Observer.	Date.	Intensity.
Slidre............	61 05	8 09	Hansteen	1821	1·454
Idsat	62 57	11 18	Hansteen	1825	1·452
Bodoe............	67 15	13 55	Keilhau	1827	1·451
Bear Island	74 55	14 50	Keilhau	1827	1·496
Spitzbergen, Whale's Head .. }	77 25	17 00	Keilhau	1827	1·539
Tromsoe	69 38	18 55	Keilhau	1827	1·515
Jacob's Elv	69 54	20 45	Keilhau	1827	1·467
‡ Talvig	70 02	22 48	Keilhau	1827	1·512
Havoe Sund	70 57	23 19	Keilhau	1827	1·476
Ingoe..........	71 06	24 03	Keilhau	1827	1·517
Mageroe	71 01	26 01	Keilhau	1827	1·500
Hammerfest	70 40	23 46	Sabine	1823	1·506
			Keilhau	1827	1·461
„	„	„	Keilhau	1827	1·461
Upper Tornea	66 16	23 47	Hansteen	1825	1·464
Brahestad	64 41	24 20	Hansteen	1825	1·455
Lebbesbye......	70 37	26 45	Keilhau	1827	1·465
Mehavn........	71 06	27 53	Keilhau	1827	1·496
§ Kaleboton	70 12	28 10	Keilhau	1827	1·491
Omgang.........	71 00	28 30	Keilhau	1827	1·487
Berlevaag	70 54	29 11	Keilhau	1827	1·460
Wadsoe........	70 10	29 50	Keilhau	1827	1·469

* Mean, 3 stations	35 41	237 06	1·583
† Mean, 3 stations	34 34	238 27	1·585
‡ Mean, 6 stations	70 26	22 38	1·498
§ Mean, 6 stations	70 40	28 23	1·478

Station.	Lat.	Long.	Observer.	Date.	Intensity.
Wardhuus	70 23	31 07	Keilhau	1827	1·477
Miteschka	56 13	49 54	Erman	1828	1·459
„	„	„	Hansteen & Due. ...	1828	1·447
Milet	56 41	50 30	Erman............	1828	1·473
„	„	„	Hansteen & Due. ...	1828	1·461
Koschil	57 08	51 52	Erman............	1828	1·488
„	„	„	Hansteen & Due. ...	1828	1·478
Suri	57 34	53 23	Erman............	1828	1·476
„	„	„	Hansteen & Due. ...	1828	1·477
Dubrowa	57 42	54 30	Erman............	1828	1·482
„	„	„	Hansteen & Due. ...	1828	1·488
Ochansk	57 00	56 00	Hansteen & Due. ...	1828	1·497
Perm	58 01	56 14	Hansteen & Due. ...	1828	1·494
„	„	„	Erman............	1828	1·489
Krilassowa........	57 34	56 37	Hansteen & Due	1828	1·501
„	„	„	Erman............	1828	1·535
Buikowa..........	56 53	57 26	Hansteen & Due. ...	1828	1·504
„	„	„	Erman............	1828	1·514
Kirgischansk	56 50	59 06	Hansteen & Due. ...	1828	1·525
„	„	„	Erman............	1828	1·509
Kushwa	58 17	59 43	Hansteen & Due. ...	1828	1·500
„	„	„	Erman............	1828	1·502
N. Tagilsk	57 55	59 54	Hansteen & Due. ...	1828	1·506
Bogoslowsk	59 49	59 55	Erman............	1828	1·524
„	„	„	Hansteen & Due. ...	1828	1·509
Ekaterinenburg. ...	56 51	60 34	Erman............	1828	1·522
„	„	„	Hansteen & Due. ...	1828	1·524
Werchoturie	58 52	60 46	Erman............	1828	1·548
„	„	„	Hansteen & Due. ...	1828	1·536
Bjelieska..........	56 50	61 56	Erman......... ;...	1828	1·509
„	„	„	Hansteen & Due. ...	1828	1·508
Sugazk	57 00	63 44	Erman............	1828	1·501
„	„	„	Hansteen & Due. ...	1828	1·535
Tiumen	57 10	65 27	Erman............	1828	1·505
„	„	„	Hansteen & Due. ...	1828	1·550
Nishnei Turinsk. ...			Hansteen & Due. ...	1828	1·535
Orlowa			Hansteen & Due. ...	1828	1·543
Semipalatinsk......	50 24	80 21	Hansteen	1829	1·556
Natschika	53 06	158 15	Erman............	1829	1·494
St. Peter and St. Paul	53 00	158 40	Erman............	1829	1·489
Kosuirewsk	55 52	159 34	Erman............	1829	1·548
Chartschinsk......	56 31	160 43	Erman............	1829	1·542
Ielowka	56 54	160 55	Erman............	1829	1·543
Kuruginski........	58 34	163 27	Lütke	1828	1·533
At Sea	40 28	213 35	Lütke	1827	1·456

Station.	Lat.	Long.	Observer.	Date.	Intensity.
Cayman Island	19 14	278 55	Sabine.............	1822	1·450
Terceira..........	38 39	332 47	Fitz Roy..........	1836	1·457
Greenland	74 32	341 10	Sabine............	1823	1·543

§ 5. *Intensities from* 1·45 *to* 1·35.

Station.	Lat.	Long.	Observer.	Date.	Intensity.
Brussels	50 52	4 20	Quetelet	1829	1·374
,, ...	,,	,,	Rudberg	1832	1·369
⎡ Bekkervig	60 01	5 10	Hansteen..........	1821	1·411
⎮ Bergen	60 24	5 17	Hansteen..........	1821	1·422
• ⎮ Ullensvang	60 20	6 38	Hansteen..........	1821	1·426
⎨ Leierdal........	61 10	7 50	Hansteen..........	1821	1·419
⎮ Mariasteen......	61 02	8 14	Hansteen..........	1821	1·406
⎣ Norsteboe	60 20	8 37	Hansteen..........	1821	1·414
Francfort	50 10	8 37	Quetelet	1829	1·358
Tubingen	48 31	9 04	Humboldt&G.Lussac	1806	1·357
⎡ Ingolfsland	59 53	8 48	Hansteen..........	1821	1·416
⎮ Bolkesjoë	59 43	9 20	Hansteen..........	1821	1·405
† ⎨ Korset	58 49	9 32	Hansteen..........	1822	1·373
⎮ Kongsberg......	59 40	9 40	Hansteen..........	1820	1·414
⎣ Helgeroe	58 59	9 54	Hansteen..........	1822	1·398
Kolding	55 27	9 20	Hansteen..........	1824	1·385
Sleswig	54 31	9 55	Hansteen..........	1824	1·381
Gottingen	51 32	9 55	Humboldt&G.Lussac	1806	1·348
,, ...	,,	,,	Quetelet	1829	1·365
,, ...	,,	,,	Rudberg	1832	1·349
Aalborg	57 03	9 56	Hansteen..........	1824	1·367
⎡ Tomlevold......	60 51	9 58	Hansteen..........	1821	1·425
⎮ Heggen	59 55	10 10	Hansteen..........	1825	1·415
⎮ Drammen	59 49	10 13	Hansteen..........	1823	1·377
‡ ⎨ Moe	60 14	10 31	Hansteen..........	1821	1·423
⎮ Gran	60 22	10 32	Hansteen..........	1821	1·422
⎣ Johnsrud	59 57	10 37	Hansteen..........	1825	1·425
Aarhuus..........	56 10	10 14	Hansteen..........	1824	1·384
Odense	55 24	10 19	Hansteen..........	1824	1·365
Drontheim........	63 26	10 25	Sabine............	1823	1·442
,, ...	,,	,,	Hansteen..........	1825	1·430
Christiania	59 55	10 45	Hansteen..........	1820	1·419

* Mean, 6 stations	60 33	7 38	1·416
† Mean, 5 stations	59 13	9 27	1·401
‡ Mean, 6 stations	60 11	10 20	1·414

Station.	Lat.	Long.	Observer.	Date.	Intensity.
Elleöen	59 19	10 40	Hansteen..........	1822	1·384
Soner..........	59 32	10 45	Hansteen..........	1822	1·383
* Skieberg	59 14	11 11	Hansteen..........	1822	1·372
Fredericshall....	59 01	11 30	Hansteen and Due..	1828	1·387
Altorp	58 53	12 14	Hansteen..........	1822	1·389
Vang..........	61 06	10 34	Hansteen..........	1821	1·431
Nebye	62 18	10 58	Hansteen..........	1825	1·423
† Biornestad......	61 03	11 28	Hansteen..........	1825	1·423
Roraas	62 34	11 35	Hansteen..........	1825	1·440
Grundsat	60 56	11 35	Hansteen..........	1825	1·440
Fredericshavn ..	57 27	10 33	Hansteen..........	1824	1·384
‡ Gottenburg	57 42	10 58	Hansteen..........	1819	1·383
Quistrum	58 27	11 45	Hansteen..........	1819	1·407
Odensala	57 26	12 03	Hansteen..........	1822	1·367
Wennersborg....	58 22	12 17	Hansteen & Due....	1828	1·381
Suul	63 42	12 12	Hansteen..........	1825	1·423
Soroe..........	55 27	11 54	Hansteen..........	1820	1·384
§ Fredericsberg ..	55 56	12 18	Hansteen..........	1820	1·403
Helsingberg	56 03	12 43	Hansteen..........	1820	1·378
Copenhagen	55 41	12 55	Hansteen..........	1820	1·367
Leipsic	51 20	12 22	Keilhau & Boeck ..	1826	1·359
"	"	"	Quetelet	1829	1·363
Magnor	59 57	12 22	Hansteen..........	1825	1·420
Berlin	52 31	13 22	Humboldt & G.Lussac	1806	1·370
"	"	"	Erman	1828	1·367
"	"	"	Quetelet	1829	1·367
Dresden..........	51 02	13 43	Quetelet	1829	1·366
Ystad	55 26	13 56	Erichsen	1824	1·374
Carlstad........	59 23	13 26	Hansteen..........	1825	1·378
‖ Mariestad	58 40	13 50	Hansteen & Due....	1828	1·381
Lincoping	58 26	15 38	Hansteen & Due....	1828	1·356
Carolath........	51 46	15 57	Erichsen	1824	1·351
Oestersund......	63 10	14 32	Hansteen..........	1825	1·434
Grimnas........	62 50	15 10	Hansteen..........	1825	1·427
¶ Alsta	62 29	16 0	Hansteen..........	1825	1·422
Sundswall	62 22	17 16	Hansteen..........	1825	1·415
Hernosand......	62 38	17 53	Hansteen..........	1825	1·421

* Mean, 5 stations	59 12	11 16	1·383
† Mean, 5 stations	61 35	11 16	1·431
‡ Mean, 5 stations	57 53	11 31	1·384
§ Mean, 4 stations	55 47	12 28	1·383
‖ Mean, 3 stations	58 50	14 18	1·372
¶ Mean, 5 stations	62 42	16 10	1·424

Station.	Lat.	Long.	Observer.	Date.	Intensity.
Geb					
Stoc					
,					
,					
Dan					
Ume					
Kön					
Tjoc					
Pite					
Was					
Bior					
Abo					
Carl					
Torr					
Ulea					
Pete					
Pomeranja	59 13	31 23	Erman............	1828	1·427
„	„	„	Hansteen & Due....	1828	1·417
G. Novgorod					
„					
Waldai					
„					
W. Wolotschok					
„					
Tver					
„					
Moscow					
„					
Platowa					
„					
Demitrewski					
„					
Murom					
„					
Osablikowo					
Doskino..........					
„					
N. Novgorod					
„					
Tschougouniei					
„					
Angikowo					
„					

Station.	Lat.	Long.	Observer.	Date.	Intensity.
Kasan.............	55 48	49 07	Erman.....·........	1828	1·440
„	„	„	Hansteen & Due....	1828	1·425
Uralsk	51 11	51 22	Hansteen..........	1829	1·398
Klinen	49 05	52 00	Hansteen..........	1829	1·370
Orenburg	51 45	55 06	Hansteen..........	1829	1·432
Oufa	54 45	56 00	Hansteen..........	1829	1·469
Havana	23 09	277 38	Humboldt	1801	1·351
„	„	„	Sabine............	1822	1·492
Jamaica	17 56	283 06	Sabine............	1822	1·436
Madeira.;.......	32 38	343 04	Sabine............	1822	1·373
„	„	„	King	1826	1·377
Ireland. By 30 } stations	53 25	352 05	Lloyd & Sabine	1835	1·410
Scotland. By 25 } stations	56 27	355 35	Sabine............	1836	1·414
Stromness	58 58	356 30	Ross..............	1836	1·419
Brassa...........;	60 09	358 48	Sabine............	1818	1·443
London	51 31	359 50	Sabine............	1827	1·372

§ 6. *Intensities from* 1·35 *to* 1·25.

Station.	Lat.	Long.	Observer.	Date.	Intensity.
⌠ Valencia	39 29	359 36	Humboldt	1798	1·241
⎟ Cambrils	40 55	0 46	Humboldt	1798	1·305
* ⎨ Barcelona	41 23	2 12	Humboldt	1798	1·348
⎟ Gerona	41 52	2 48	Humboldt	1798	1·209
⌡ Perpignan......	42 43	2 57	Humboldt	1798	1·381
Paris	48 52	2 21	Humboldt	1800	1·348
⌠ Montpellier	43 36	3 53	Humboldt	1798	1·348
† ⎨ Nismes	43 50	4 20	Humboldt	1798	1·294
⌡ Marseilles......	43 18	5 23	Humboldt	1798	1·294
⌠ Lyons	45 46	4 52	Humboldt & G. Lussac	1805	1·333
‡ ⎨ St. Michel......	45 23		Humboldt & G. Lussac	1805	1·349
⌡ M. Cenis	45 14	6 55	Humboldt & G. Lussac	1805	1·344
⌠ Geneva	46 12	6 07	Quetelet	1830	1·292
§ ⎨ Gd. St. Bernard..	45 55	7 11	Quetelet	1830	1·294
Lanslebourg	45 18		Humboldt & G. Lussac	1805	1·323

*	Mean, 5 stations	41	16	1	39	1·296
†	Mean, 3 stations	43	35	4	32	1·312
‡	Mean, 3 stations	45	28	5	53	1·342
§	Mean, 2 stations	46	03	6	39	1·293

Intensity.	Lat.	Long.	Observer.	Date.	Intensity.
Turin	45 04	7 42	Humboldt & G. Lussac	1805	1·336
● ⎰ St. Gothard	46 32	8 33	Humboldt & G. Lussac	1805	1·314
⎱ Altorp	46 41	8 32	Humboldt & G. Lussac	1805	1·325
Como	45 48	9 06	Humboldt & G. Lussac	1805	1·310
Milan	45 28	9 09	Humboldt & G. Lussac	1805	1·312
,, ...	,,	,,	Quetelet	1830	1·294
Florence..........	43 46	11 15	Humboldt & G. Lussac	1805	1·278
Munich ,.........	48 08	11 34	Erman	1826	1·339
Rome	41 54	12 26	Humboldt & G. Lussac	1805	1·264
Toplitz	49 58	12 52	Keilhau & Boeck ..	1826	1·334
† ⎰ Trieste	45 38	13 47	Keilhau & Boeck ..	1826	1·317
⎱ Lohitsch	45 55	14 13	Keilhau & Boeck ..	1826	1·314
Naples	40 50	14 14	Humboldt & G. Lussac	1805	1·274
Prague	50 05	14 27	Keilhau	1826	1·332
Gratz..........	47 04	15 27	Keilhau & Boeck	1826	1·327
‡ ⎰ Iglau	49 23	15 36	Keilhau & Boeck	1826	1·319
⎱ Vienna	48 13	16 23	Keilhau & Boeck	1826	1·325
Nicolaieff	46 58	32 01	Kupffer	1829	1·275
Taganrog	47 12	38 58	Kupffer	1829	1·308
Stavropol	45 03	42 01	Kupffer	1829	1·327
Bridge of Malka ..	43 45	42 30	Kupffer	1829	1·302
Astrachan	46 20	48 00	Hansteen..........	1830	1·334
At Sea	13 39	311 50	Humboldt	1799	1·256
At Sea	20 41	335 08	Humboldt	1799	1·256
Teneriffe..........	28 27	343 45	Humboldt	1798	1·272
,, ...	,,	,,	Freycinet..........	1817	1·340
,, ...	,,	,,	Sabine	1822	1·313
Ferrol	43 29	351 46	Humboldt	1799	1·262
Villa el Pando ..	41 58	354 33	Humboldt	1799	1·294
§ ⎰ Medina del Campo	41 24	355 16	Humboldt	1799	1·294
⎱ Guadarama	40 39	355 52	Humboldt	1799	1·294
Villa Franca	42 37	355 59	Humboldt	1799	1·294
Madrid	40 25	356 19	Humboldt	1799	1·294

● Mean, 4 stations	46 00	8 28	1·321
† Mean, 2 stations	45 45	14 00	1·315
‡ Mean, 3 stations	48 13	15 49	1·324
§ Mean, 6 stations	41 45	354 58	1·290

§ 7. *Intensities from* 1·25 *to* 1·15.

Station.	Lat.	Long.	Observer.	Date.	Intensity.
Port William	37 00	38 00	Estcourt	1836	1·198
Bussora	30 20	47 36	Estcourt	1836	1·175
At Sea	39 07	159 03	Lütke	1827	1·186
Carthagena.......	10 25	285 31	Humboldt	1801	1·294
* { Mompox	9 14	285 34	Humboldt	1801	1·199
Morales	8 15	286 0	Humboldt	1801	1·188
⌠ Nueva Valencia..	10 10	291 47	Humboldt	1800	1·197
Hac. de Cura....	10 16	292 06	Humboldt	1800	1·189
Victoria........	10 14	292 30	Humboldt	1800	1·251
† { Hac. de Tui	10 17	292 34	Humboldt	1800	1·168
Venta di Avila ..	10 33	292 53	Humboldt	1800	1·230
La Guayra.....	10 36	292 54	Humboldt	1800	1·262
Caracas	10·31	292 56	Humboldt	1800	1·209
⌊ Silla de Curacas..	10 31	292 59	Humboldt	1800	1·189
⌠ Cumana........	10 28	295 51	Humboldt	1800	1·178
Il Impossibile ..	10 26	295 55	Humboldt	1800	1·219
‡ { Cocollar........	10 10	296 01	Humboldt	1800	1·178
Caripe	10 10	296 07	Humboldt	1800	1·178
⌊ Cumanaçoa	10 16	296 02	Humboldt	1800	1·168
Trinidad.........	10 39	298 25	Sabine	1822	1·198
At Sea	10 53	299 29	Humboldt	1799	1·220
Port Praya.......	14 54	336 30	Sabine	1822	1·193
"	"	"	King	1826	1·177
"	"	"	Fitz Roy .... {	1832 1836	1·156

§ 8. *Intensities from* 1·15 *to* 1·05.

Bonin...........	27 07	142 24	Lütke	1828	1·111
Oahu	21 18	202 0	Douglas	1830	1·119
Mowi...........	20 52	203 19	Freycinet..........	1819	1·133
Owhyhee	19 43	203 50	Douglas	1834	1·098
Galapagos I.	0 15S.	269 29	Fitz Roy	1835	1·069
⌠ Guajaquil	2 13S.	280 03	Humboldt	1803	1·058
§ { Cuenca	2 55S.	280 47	Humboldt ..,....	1802	1·029
Alausi	2 13S.	281 0	Humboldt	1802	1·058
⌊ Riobamba......	1 42S.	281 16	Humboldt	1802	1·077

* Mean, 3 stations	9 18	285 42	1·227	
+ Mean, 8 stations	10 24	292 35	1·203	
‡ Mean, 5 stations	10 18	296 00	1·184	
§ Mean, 4 stations	2 16 S.	280 46	1·055	

Station.	Lat.	Long.	Observer.	Date.	Intensity.
* { Quito.........	0° 14 S.	281° 16	Humboldt	1802	1·067
San Antonio	0 0	281 19	Humboldt	1802	1·087
Villa di Ibarra ..	0 21	281 42	Humboldt	1802	1·028
† { Pasto..........	1 13	282 39	Humboldt	1801	1·048
Almaquer	1 54	283 06	Humboldt	1801	1·067
Popoyan	2 38	283 21	Humboldt	1801	1·117
‡ { Carthago	4 45	283 54	Humboldt	1801	1·077
Ibague	4 27	284 41	Humboldt	1801	1·147
S. Fé de Bogota	4 36	285 47	Humboldt	1801	1·147
§ { Honda........	5 12	285 07	Humboldt	1801	1·117
Bocca di Nares .	6 10	285 20	Humboldt	1801	1·137
‖ { Atabapo	4 03	291 50	Humboldt	1800	1·077
Apure	7 53	292 01	Humboldt	1800	1·107
Atures	5 38	292 02	Humboldt	1800	1·117
Carichana	6 34	292 06	Humboldt	1800	1·157
Calabozo	8 56	292 10	Humboldt	1800	1·107
¶ { Iavita	2 48	291 59	Humboldt	1800	1·068
St. Carlos	1 54	292 22	Humboldt	1800	1·048
** { Nueva Barcelona	10 07	295 16	Humboldt	1800	1·127
St. Thomas ...	8 08	296 06	Humboldt	1800	1·107
River Gambia	13 08	343 27	Sabine...........	1822	1·141
Sierra Leone	8 29	346 45	Sabine...........	1822	1.053

§ 9. *Intensities from* 1·05 *to* 0·95.

Manilla	14 36 N.	116 18	Lütke	1829	1·044
Guahan	13 26	144 44	Lütke	1829	0·980
Agagna	13 28	144 58	Freycinet........	1818	0·968
At Sea	6 55	158 02	Lütke	1827	0·990
At Sea	11 27	161 52	Lütke	1827	0·970
At Sea	2 56	162 50	Lütke	1827	1·018
At Sea	4 17	162 54	Lütke	1827	1·001
At Sea	3 47	162 59	Lütke	1827	1·010
At Sea	18 44	163 55	Lütke	1827	0·989
At Sea	0 35 N.	232 56	Lütke	1827	1·013

* Mean, 3 stations	0° 2	281° 26	1·061
† Mean, 3 stations	1 52	283 02	1·077
‡ Mean, 3 stations	4 36	284 47	1·124
§ Mean, 2 stations	5 31	285 14	1·127
‖ Mean, 5 stations	6 36	292 02	1·113
¶ Mean, 2 stations	2 21	292 10	1·058
** Mean, 2 stations	9 07	295 41	1·117

Station.	Lat.	Long.	Observer.	Date.	Intensity.
⎧ Ayavaca	4° 38 S.	280° 26	Humboldt	1802	1·019
⎪ Gualtaquillo	4 52 S.	280 26	Humboldt	1802	1·028
⎪ Gonzanama	4 13 S.	280 27	Humboldt	1802	1·009
*⎨ Guancabamba ..	5 14 S.	280 37	Humboldt	1802	1·019
⎪ Pucara	5 56 S.	280 37	Humboldt	1802	1·009
⎪ Amazon's River..	5 48 S.	281 13	Humboldt	1802	1·009
⎩ Tomependa	5 31 S	281 24	Humboldt	1802	1·019
⎧ Montan	6 33 S.	281 10	Humboldt	1802	1·009
†⎨ Micuipampa	6 44 S.	281 21	Humboldt	1802	1·000
⎪ Santa	8 59 S.	281 23	Humboldt	1802	1·019
⎩ Caxamarca.	7 09 S.	281 25	Humboldt	1802	1·019
Maranham	2 32 S.	315 39	Sabine	1822	1·016

§ 10. *Intensities below 0·95.*

Station.	Lat.	Long.	Observer.	Date.	Intensity.
St. Thomas.......	0 25	6 45	Sabine	1822	0·931
St. Catherine......	27 26 S.	311 27	King	1827	0·920
Rio de Janeiro	22 55 S.	316 51	Freycinet. .... ⎰	1817 1820	0·890
„	„	„	Lütke	1827	0·886
„	„	„	Erman............	1830	0·879
„	„	„	Fitz Roy	1832	0·878
Bahia	12 59 S.	321 30	Sabine	1822	0·898
„	„	„	Fitz Roy	1836	0·871
Pernambuco	8 04 S.	325 09	Fitz Roy	1836	0·914
Ascension	7 56 S.	345 36	Sabine	1822	0·920
„	„	„	Fitz Roy	1836	0·873
St. Helena........	15 55 S.	354 17	Fitz Roy	1836	0·836

DIVISION II.   SOUTHERN HEMISPHERE.

§ 11. *Intensities from 0·95 to 1·05.*

Station.	Lat.	Long.	Observer.	Date.	Intensity.
Cape of Good Hope	34 11 S.	18 26	Freycinet.	1818	0·945
„	„	„	Fitz Roy	1836	1·014
Rawak	1 34 S.	131 00	Freycinet.	1818	1·044
Ulean............	7 22 N.	143 57	Lütke	1828	1·004
Lugunor..........	5 29 N.	153 58	Lütke	1828	0·998
Los Valientes......	5 46 N.	157 05	Lütke	1828	0·993

* Mean, 7 stations   5° 10 S.   280° 43′   1·017
† Mean, 4 stations   7 21 S.   281 20   1·012

Station.	Lat.	Long.	Observer.	Date.	Intensity.
Ualan............	5° 21 N.	163° 23'	Lütke	1827	1·002
At Sea	4 20 S.	238 13	Lütke	1827	0·998
At Sea	13 09 S.	251 20	Lütke	1827	1·014
⎰ Casma	9 38 S.	281 25	Humboldt	1802	1·000
⎪ Guarmey	10 04 S.	281 39	Humboldt	1802	1·000
*⎨ Huaura	11 03 S.	282 14	Humboldt	1802	1·009
⎪ El Ramadal	11 32 S.	282 35	Humboldt	1802	1·009
⎱ Lima..........	12 03 S.	282 53	Humboldt	1802	1·077
Goriti	34 57 S.	305 03	King	1829	1·041

§ 12. Intensities from 1·05 to 1·15.

Station.	Lat.	Long.	Observer.	Date.	Intensity.
Mauritius	20 09 S.	57 31	Freycinet..........	1818	1·096
„	„	„	Fitz Roy	1836	1·192
Amboyna	3 42 S.	128 08	Rossel	1792	1·097
Otaheite..........	17 29 S.	210 30	Erman............	1830	1·172
„	„	„	Fitz Roy	1835	1·017
Coquimbo	29 59 S.	288 34	Fitz Roy	1835	1·111
Blanco Bay	38 57 S.	298 01	Fitz Roy	1832	1·113
Monte Video	34 53 S.	303 47	King	1830	1·065
„	„	„	Fitz Roy	1833	1·055
At Sea	40 55 S.	307 00	Lütke	1827	1·110

§ 13. Intensities from 1·15 to 1·25.

Station.	Lat.	Long.	Observer.	Date.	Intensity.
Timor............	10 10 S.	123 40	Freycinet..........	1818	1·177
Valdivia..........	39 53 S.	286 31	Fitz Roy	1835	1·238
Concepcion	36 42 S.	286 50	Lütke	1827	1·234
„	„	„	King	1829	1·250
„	„	„	Fitz Roy	1835	1·186
Valparaiso	33 02 S.	288 19	Lütke	1827	1·170
„	„	„	King .... ⎰ 1829 ⎱ 1830		1·176

§ 14. Intensities from 1·25 to 1·35.

Station.	Lat.	Long.	Observer.	Date.	Intensity.
Juan Fernandez....	33 38 S.	281 07	King	1830	1·262
At Sea	41 00 S.	282 30	Lütke	1827	1·324
Port Low	43 48 S.	285 58	Fitz Roy..........	1835	1·326
Chiloe............	41 51 S.	286 04	King	1829	1·321
„	„	„	Fitz Roy..........	1834	1·304
At Sea	49 18 S.	302 48	Lütke	1827	1·268

* Mean, 5 stations   10° 52' S.   282° 10'   1·019.

§ 15. *Intensities from* 1·35 *to* 1·45.

Station.	Lat.	Long.	Observer.	Date.	Intensity.
Bay of Seals	25 43 S.	113 20	Freycinet,.........	1818	1·421
R. Santa Cruz	50 07 S.	291 36	Fitz Roy	1834	1·425
Port Desire	47 45 S.	294 05	Fitz Roy	1833	1·355
Sea Bear Bay......	47 51 S.	294 12	King	1829	1·361
At Sea	55 25 S.	298 27	Lütke	1827	1·413
Falkland Ids........	51 33 S.	301 55	Freycinet..........	1820	1·363
,, ...	51 32 S.	301 53	Fitz Roy	1833	1·349
,, ...	,,	,,	Fitz Roy	1834	1·385

§ 16. *Intensities from* 1·45 *to* 1·55.

Station.	Lat.	Long.	Observer.	Date.	Intensity.
Port Famine	53 38 S.	289 02	King	1827	1·505
,, ...	,,	,,	Fitz Roy	1834	1·560
St. Martin's Cove ..	55 51 S.	292 26	King	1827	1·498

§ 17. *Intensities from* 1·55 *to* 1·65.

Station.	Lat.	Long.	Observer.	Date.	Intensity.
New Zealand	35 16 S.	174 00	Fitz Roy	1835	1·591

§ 18. *Intensities from* 1·65 *to* 1·75.

Station.	Lat.	Long.	Observer.	Date.	Intensity.
Sydney	33 51 S.	151 17	Freycinet..........	1819	1·631
,, ...	,,	,,	Fitz Roy	1836	1·685
King George's Sound	35 02 S.	117 56	Fitz Roy	1836	1·709

§ 19. *Intensities from* 1·75 *to* 1·85.

Station.	Lat.	Long.	Observer.	Date.	Intensity.
Hobart Town......	42 53 S.	147 24	Fitz Roy	1836	1·817

Additional Table, containing the Observations made by M. Erman at sea, on his return from Kamtschatka to Europe by Cape Horn.

These observations were received from M. Erman since this Report was sent to press, which occasions their being given in a separate table.

	Latitude.	Longitude.	Dip.	Intens.
Pacific Ocean	51 03	203 52	67 09·5 N.	1·522
	53 35	213 38	71 05·5	1·587
	55 33	221 01	75 39·1	1·639
„	54 27	221 23	73 40·0	1·673
	43 18	230 24	66 44·5	1·580
„	40 03	233 39	64 00·7	1·551
	39 12	235 28	63 40·0	1·528
„	38 0	235 54	63 41·5	1·556
	31 51	234 18	56 31·9	1·435
„	30 31	235 41	55 05·0	1·394
	29 04	238 24	53 20·8	1·380
	28 41	238 59	53 05·5	1·402
„	28 04	239 08	52 09·5	1·364
	26 36	239 28	50 22·6	1·377
„	26 0	238 54	49 26·1	1·321
	25 21	238 37	48 06·5	1·356
	23 12	238 15	45 20·5	1·341
„	23 0	238 12	44 14·4	1·289
	21 14	237 57	42 17·0	1·271
	19 39	237 45	40 07·8	1·241
„	18 36	237 34	39 03·0	1·219
	16 56	237 13	35 34·7	1·185
„	15 15	236 55	32 28·4	1·183
	13 37	236 36	29 45·7	1·158
„	12 18	236 28	27 09·3	1·143
	11 18	236 22	25 44·4	1·136
	9 43	235 58	23 06·4	1·107
„	8 55	235 57	20 57·7	1·082
	7 15	236 26	17 51·9	1·053
	6 27	236 42	17 08·8	1·055
„	5 49	236 38	15 24·8	1·056
	4 35	235 47	13 02·6	1·049
„	2 42	234 17	9 18·0	1·028
	1 33	233 29	7 21·2	1·018
	0 46	232 54	5 15·4	0·992
„	0 9	232 27	3 30·4	0·986

	Latitude.	Longitude.	Dip.	Intensity.
Pacific Ocean	0 12 S.	232 09	3 8·5	0·997
	0 6 S.	231 44		0·995
	0 7 N.	230 40	3 45·3	1·014
	0 8 N.	229 44	4 19·3	1·022
"	0 0 N.	229 22	3 49·5	1·029
	0 29 S.	228 41	2 38·3	0·977
"	0 40 S.	228 30	2 16·8	0·980
·	0 53 S.	228 16	2 10·9	1·000
"	1 7 S.	228 0	1 32·8	1·028
	1 47 S.	227 18	0 14·6 S.	1·015
"	1 52 S.	226 28	0 16·2 N.	0·996
	1 53 S.	225 32	0 42·6 S.	0·942
"	1 52 S.	225 08	0 0·9 N.	1·008
	1 30 S.	223 46	0 46·7 N.	1·015
"	1 37 S.	222 12	0 57·4 N.	1·004
	1 48 S.	221 49	0 3·7 S.	1·009
	2 11 S.	221 13	0 21·8 S.	1·022
"	1 57 S.	221 0		1·001
	2 19 S.	220 16	0 39·4 S.	0·981
	4 30 S.	218 42	5 3·9 S.	1·016
"	5 34 S.	218 3	7 29·8 S.	1·032
	7 08 S.	217 4	10 07·3 S.	1·031
"	7 45 S.	216 53	11 27·1 S.	1·009
	8 06 S.	216 41	12 46·8 S.	1·033
	9 22 S.	215 58	15 18·5 S.	1·066
"	10 22 S.	215 21	17 16·7 S.	1·105
	11 13 S.	214 59	18 18·0 S.	1·081
"	11 54 S.	214 52	19 10·8 S.	1·070
	12 2 S.	214 51	19 32·9 S.	1·114
"	12 56 S.	214 38	21 19·1 S.	1·118
	13 7 S.	214 37	21 16·9 S.	1·124
"	13 44 S.	214 51	22 23·6 S.	1·095
	14 01 S.	214 31	23 28·6 S.	1·075
	14 55 S.	213 59	24 54·2 S.	1·121
"	14 49 S.	212 26	24 23·2 S.	1·091
	19 06 S.	209 49	31 56·5 S.	1·253
"	22 17 S.	209 29	35 51·8 S.	1·209
·	24 51 S.	210 0	40 19·4 S.	1·250
	26 56 S.	209 54	43 05·5 S.	1·349
"	27 43 S.	209 57	44 02·9 S.	1·324
	28 48 S.	213 08	45 27·9 S.	1·257
	29 04 S.	213 25	45 26·5 S.	1·339
"	30 33 S.	212 58	47 20·6 S.	1·371

	Latitude.	Longitude.	Dip.	Intensity.
Pacific Ocean	32 22 S.	214 35	49 07·1 S.	1·361
,,	34 23 S.	216 27	51 12·7 S.	1·370
,,	34 55 S.	218 29	52 29·3 S.	1·392
,,	34 28 S.	220 19	50 32·9 S.	1·426
,,	36 17 S.	219 50	52 17·6 S.	1·407
,,	37 39 S.	218 4	53 52·4 S.	1·489
,,	42 04 S.	218 44	58 48·4 S.	1·509
,,	44 24 S.	221 59	61 4·2 S.	1·543
,,	45 6 S.	225 11	61 56·7 S.	1·545
,,	45 05 S.	228 23	61 43·9 S.	1·611
,,	47 13 S.	237 34	63 15·5 S.	1·583
,,	48 11 S.	242 23	63 39·6 S.	1·609
,,	48 50 S.	245 29	64 25·5 S.	1·666
,,	51 03 S.	252 22	65 48·6 S.	1·614
,,	55 03 S.	266 24	66 16·1 S.	1·630
,,	56 28 S.	276 38	65 05·6 S.	1·576
,,	56 05 S.	284 36	62 51·3 S.	1·537
,,	58 31 S.	289 35	61 05·6 S.	1·522
Atlantic Ocean	57 26 S.	295 56	60 06·5 S.	1·491
,,	56 02 S.	299 34	58 26·6 S.	1·391
,,	55 36 S.	302 02	57 28·4 S.	1·412
,,	52 44 S.	304 26	54 29·0 S.	1·301
,,	50 12 S.	304 ·17	51 09·5 S.	1·280
,,	47 11 S.	306 20	48 44·5 S.	1·233
,,	39 48 S.	308 45	40 27·0 S.	1·023
,,	37 09 S.	309 41	36 41·9 S.	1·016
,,	35 44 S.	310 23	34 09·9 S.	0·938
,,	33 04 S.	312 02	30 3·4 S.	0·984
,,	29 53 S.	312 28	25 32·5 S.	0·923
,,	27 58 S.	314 20	22 01·2 S.	0·899
,,	26 22 S.	315 30	19 44·7 S.	0·880
,,	24 12 S.	316 19	16 02·0 S.	0·844
,,	24 24 S.	316 12	15 47·9 S.	0·916
,,	24 18 S.	318 35	16 35·0 S.	0·867
,,	24 53 S.	324 26	18 29·9 S.	0·852
,,	24 26 S.	325 12	15 17·1 S.	0·811
,,	24 06 S.	325 14	15 56·6 S.	0·809
,,	20 56 S.	325 15	9 45·1 S.	0·816
,,	20 00 S.	325 0	7 53·3 S.	0·743
,,	19 38 S.	324 56	7 34·0 S.	0·792
,,	18 57 S.	324 57	7 19·8 S.	0·820
,,	17 33 S.	325 54	4 44·0 S.	0·784
,,	16 17 S.	325 30	2 28·0 S.	0·795
,,	15 56 S.	326 33	1 33·5 S.	0·797
,,	14 53 S.	326 49	0 24·8 N.	0·838

	Latitude.	Longitude.	Dip.	Intensity.
Atlantic Ocean .... {	14 25 S.	327 05	1 28·8 N.	0·856
	13 18 S.	327 22	3 18·2	0·812
,,	9 42 S.	328 15	9 28·0	0·892
,,	5 19 S.	329 12	17 43·0	0·922
	3 51 S.	329 19	20 24·2	0·949
,, {	1 53 S.	329 33	23 28·9	1·031
,,	0 26 N.	329 45	27 16·5	1·043
,,	2 30	329 32	30 48·4	1·074
,,	4 26	329 56	34 29·5	1·094
,,	5 45	331 21	35 16·5	1·094
	9 36	333 34	39 14·4	1·125
,, {	10 24	333 35	40 48·3	1·114
	11 3	332 38	41 54·8	1·187
,, {	12 36	331 42	44 4·3	1·209
	14 36	330 58	46 20·9	1·201
	15 53	329 26	48 15·9	1·273
,,	16 41	328 48	49 52·0	1·238
	19 05	326 42	51 59·6	1·311
,, {	21 01	325 07	54 44·0	1·314
,,	24 0	322 55	58 17·2	1·375
	26 26	321 55	60 49·0	1·406
,, {	28 02	321 22	61 53·6	1·404
	29 34	320 14	63 12·0	1·427
,,	30 30	319 29	64 17·3	1·478
	31 11	320 12	64 45·7	1·469
,, {	32 55	319 3	65 21·3	1·468
	33 45	318 36	66 4·4	1·499
,,	34 29	318 18	67 26·5	1·500
	35 0	318 33	67 36·6	1·505
,, {	36 15	319 56	68 17·5	1·507
	37 26	321 22	68 19·4	1·501
,, {	38 24	322 57	69 07·4	1·491
	40 09	325 20	69 32·9	1·504
,, {	41 27	327 25	70 03·6	1·466
,,	42 29	328 34	69 47·6	1·512
,,	44 22	330 55	71 07·1	1·515
,,	46 46	335 42	70 18·5	1·463
	47 47	343 58	69 46·0	1·421
,, {	47 46	344 25	70 14·9	1·419
	48 13	347 7	69 27·8	1·422
,,	49 16	351 58	69 10·5	1·416
British Channel	50 48	358 54	68 45·0	1·380

Section III.—General Conclusions.

In considering the comparative fitness of the three kinds of magnetic lines, those of equal variation, equal dip, and equal intensity, to promote a knowledge of the system of terrestrial magnetism, the lines of equal intensity have in one leading respect an advantage over the other two. Viewed under the most favourable circumstances and in its simplest aspect, the magnetism of the earth is still, it must be acknowledged, a highly complicated subject; and needs not the additional complication of its phænomena being involved with considerations foreign to itself. Now the lines of equal dip and equal variation do not express simple magnetic relations. The lines of equal dip, for example, connect those stations on the earth's surface where the direction of the magnetic attraction forms a certain angle with the horizontal plane at the station. But every station has its own horizontal plane depending on the direction of gravity, which has no known or necessary connexion with magnetism. The zero planes thus differing, the equality of dip does not express, or necessarily imply, a simple magnetic relation, but has reference to the attraction of gravitation as well as to that of magnetism. The lines of equal variation express a complex relation of a similar character. Here also the zero planes change with the station; and, the variation being the same at two stations, by no means implies parallelism in the direction of the needle at them, or any other specific relation whatsoever independent of the geographical pole, which pole has no known or necessary connexion with magnetism. It is not the same with the lines of equal intensity. Whatever may be the sources of magnetic attraction, and wherever their situation in space,—whether superficial as regards the earth,—or above or beneath its surface—the line of equal intensity expresses the equality of their resultant at all those points of the earth's surface through which it is drawn, unmixed with any considerations foreign to magnetism. They are pure magnetical isodynamic lines at the surface of the globe; and express a common relationship to the sources of magnetical attraction. The instruction they convey is therefore more simple, direct and unequivocal than in the case of the other two. The eye of the mathematician may discern the pure magnetic indication through the complex signification of the lines of equal variation and dip; but the lines of intensity are better suited to convey the system of magnetism as indicated by the phænomena to the general apprehension.

I proceed to notice a few of the most striking inferences which are deducible from the observations of intensity recorded in this report.

1. *The lines of equal intensity are not parallel with the lines of equal dip, and the difference is systematic.*

In 1805 M. Biot published an investigation of the laws which should govern the dip and the intensity, in the hypothesis of a magnet situated at the centre of the earth, having its poles infinitely near to each other, and directed to opposite points on the surface of the globe. It is a well-known consequence of this hypothesis, that the lines of equal dip and equal intensity on the earth's surface should everywhere be parallel to each other.

It has always appeared to me that the distinguished author of this investigation has been taken much beyond his meaning, when he has been supposed to have propounded this hypothesis as a general representation of the facts of terrestrial magnetism then known, or of those which should be shown by more extensive experience. He was doubtless fully aware that, many years antecedently, the phænomena of the variation had been shown by Dr. Halley to be wholly irreconcileable with the geometrical deductions from a single central magnetic axis; and that Euler, who may in some degree be regarded as an opponent of Halley upon the subject generally, fully acquiesced in this conclusion. Accordingly, M. Biot made no comparison of the hypothesis with the variation, considering no doubt that its inapplicability in that respect had been already shown. A few facts of the dip were the only observations with which he compared the formulæ of his hypothesis, and with some of these it appeared to accord tolerably; but still there were anomalies which drew from him the acknowledgement, that to represent even those few facts of the dip, it would be necessary to add to the influence of the primary axis the supposition of subordinate centres. That he had no expectation of its proving applicable to the intensity, any more than to the variation, is, I think, beyond a question, when we read the following sentence: "Quant à la declinaison et à l'intensité nous avouons franchement que nous ne savons absolument rien sur leurs lois ni sur leurs causes : et si quelque physicien est assez heureux pour les ramener à un principe unique, qui explique en même temps les variations de l'inclinaison, ce sera sans doute une des plus belles découvertes que l'on ait jamais faites."[*]

* *Journal de Physique*, vol. lix. p. 450. The state in which the question was left by Halley and Euler was, I believe, as follows : Halley decided in

The light in which I have thus considered M. Biot's essay is the same, I think, in which it was regarded at the time, by his distinguished coadjutors in this and so many other branches of science. MM. Gay Lussac and Humboldt, in closing the account of their magnetic observations on the continent of Europe in 1805 and 1806, remark as follows : " Les inclinaisons correspondantes données par la théorie d'après M. Biot, sont toutes beaucoup plus grandes, car les plus petites différences vont à près de 4°. En supposant la position de l'équateur magnétique, rigoreusement déterminée, il en resulteroit qu'en Europe, il y a une inflexion considérable des parallèles magnétiques vers l'équateur, occasionnée par l'influence de quelque centre particulier. Mais pour tirer aucune conclusion à cet égard, il est prudent d'attendre que des observations exactes et plus nombreuses fournissent *des bases solides, sur lesquelles on puisse élever une théorie rigoreuse qui les embrasse toutes**." It is here fully recognised that M. Biot's was not " cette théorie rigoreuse" which, resting on the solid basis of induction from a competent assemblage of facts, should have a proportionate claim to be regarded as a general representation of the phænomena.

In showing the incompatibility with subsequent observations of this " abstraction mathématique," as M. Biot himself designated it, I do not therefore consider myself as opposing either his opinions or his expectations.

It has sometimes appeared to me that the very simplicity of the laws of this hypothesis has tended to counterbalance in some degree the advantage it produced, in recalling attention to a subject, the interest in which had been for some years suspended. Apart from the question of accordance or non-

favour of four poles, as the best representation of the phænomena: Euler hesitated to accede to this until it should be shown more decisively that the phænomena might not be represented by a single excentric axis, having its semi-axes of unequal length; claiming in such case the preference for the latter supposition over that of four poles, as being more suitable for geometrical deductions. To have accomplished what such men as Halley and Euler had left incomplete would have been an undertaking not unworthy of M. Biot; but it would have required the preliminary labour of collecting together, as M. Hansteen has since done, the great body of the facts of observation, which, at the time his essay was written, were scattered in the journals of travellers and navigators, and in the transactions of learned societies of many countries. This labour might well in prospect have deterred him from the attempt ; but it was indispensable for the purpose of furnishing the basis of a philosophical induction of such general laws as should comprehend the whole of the phænomena. On no less solid foundation was it probable that phænomena should be represented, known to wear so complicated an aspect, and which had been the subject of the long-continued investigation of the eminent men above noticed.

* I have put in Italics the part of this extract to which I particularly refer.

accordance with facts, simplicity recommends itself to all; and persons imperfectly acquainted with the phænomena may have been led by it to undervalue *observation*, when detached portions of its facts, inconsistent with the hypothesis, may have come under their notice; and, departing from the principles of inductive philosophy, may have suffered themselves to look to the hypothesis rather than to the phænomena.   The simplicity of its resulting phænomena is, however, that characteristic in which it specially departs from the facts of nature.   The real phænomena are complex, as all who have studied them will most readily admit; and it can scarcely be expected that the laws which are to represent them should not also have in some degree an appearance of complexity, until the laws of their causation shall be discovered.

In a science which stands in need of national aid for its experimental extension, it is peculiarly desirable to remove such erroneous impressions as militate against a belief in the value, and consequently the importance, of experimental research.

I propose, therefore, in the first place, to show, that the irreconcilability of a single central axis does not rest on insulated facts only, or, as some may have supposed, on the conclusions of a single observer, but that all those who have principally concurred in extending the boundaries of our experimental knowledge of late years, have arrived at the same conclusion in that respect, and have uniformly borne testimony to the inapplicability of the formulæ of that hypothesis to represent their respective observations; and, secondly, to direct the reader's attention to those facts in particular, which may produce the readiest conviction of the systematic departure of the lines of dip and intensity from that law of the hypothesis by which they should have parallel courses.

We have already seen the conclusion at which MM. Gay Lussac and Humboldt arrived in 1807, namely, that their observations in France, Italy, and Germany, taken in conjunction with M. de Humboldt's in America, could only be reconciled with M. Biot's hypothesis, by supposing the existence of a secondary centre extending its influence over the continent of Europe, and acting conjointly with the primary.

From 1807 the spirit of experimental inquiry slumbered for a while; the times were unpropitious to a research which required freedom of access to different countries, and safety and facility in traversing extensive spaces of the earth's surface. At length it revived nearly simultaneously, in Capt. de Freycinet's voyage of circumnavigation, and in the British expeditions for the discovery of a north-west passage.   Between 1818 and

1823 I had the good fortune to enjoy opportunities of observing the magnetic phænomena over a portion of the globe amounting to about one-eighth of its surface, or the quarter of an hemisphere. In comparing, on my return to England, the observations of dip with M. Biot's formula, the differences between calculation and experiment were seen to be not at single stations only, but *systematic*, extending over large spaces of the globe; the discrepancies were also so great as (in the words which I employed in 1825) to make it " certain that no two positions could " be assigned to the magnetic poles, which would enable a cal- " culation of the dip as a function of the magnetic polar distance, " in which differences from fact should not be found of 10° and " upwards." Further, in comparing the observations of dip and intensity with the parallel course, which, according to the hypothesis, the lines of equal dip and equal intensity should preserve, their irreconcilability with this law was shown to be so great and so systematic as to be " decisive against the sup- " posed relation of the force to the observed dip; and equally " so against any other relation whatsoever, in which the re- " spective phænomena might be supposed to vary in corre- " spondence with each other." Another important difference was also pointed out. In the hypothesis the maxima of dip and intensity are coincident: with this the observations were at variance; those of the intensity placing its maximum several degrees to the southward of the geographical position which the observations of dip indicated as that of the dip of 90 degrees[*].

In 1830 M. Erman returned from a journey in which he had carried magnetic observations over a space on the globe still more extensive than mine, and (which should be specially noticed) so entirely distinct from mine, that we had not a single

* The observations of intensity arranged around their own centre presented much less discordance with the laws of an uniaxal hypothesis than appeared in those of the dip when referred to the position of the pole as indicated by the dip of 90 degrees. By substituting in the formula of that hypothesis the "itinerary distance from the maximum of intensity" for the "magnetic polar distance," and employing this formula as an empirical representation, it was found to correspond with the facts of the intensity within the district comprised by my observations, with no very material discrepancies. In that portion of the hemisphere in which the influence of the primary centre is predominant, the variations of the intensity may be easily imagined not to differ greatly from the effect of a single axis; and such is apparently the fact. It happened that my observations, extensive as they were, fell within that limit; had they been pursued a few degrees further to the eastward, the influence of the Siberian centre would have become more sensible, and the uniaxal formula would have ceased to afford even an approximate representation of the facts. But this perhaps will be better understood when the sequel of the report has been read.

station in common. I cannot state his conclusions better than
by giving his own words*.

 "*Lignes à égale Intensité, ou Lignes Isodynamiques.*—
Esperant encore completer mes observations relativement à
ces lignes interessantes, pendant mon passage du Brésil en
Europe, je me borne ici à en relever quelques particularités frap-
pantes, et nommement celle, qu'en Siberie les lignes isodynamiques ne sont rien moins que parallèles aux lignes d'égale
inclinaison. Nous voyons au contraire sous le meridien d'Ob-
dorsk et de Tobolsk, les premières avoir des branches déscen-
dantes presque verticales ou légèrement infléchies du N.O.
au S.E., tandis que les lignes à inclinaison égale y sont presque
horizontales.

　　　*　　　*　　　*　　　*　　　*

 " Ces indications préliminaires suffiront pour prouver que
l'ancienne théorie, développée par Euler et Krafft, et plus
tard par MM. Humboldt et Biot, et qui ne suppose *qu'un seul
axe magnétique*, est absolument en défaut pour les loix de l'intensité de la force magnétique. En effet, l'intensité n'étant
d'après cette théorie qu'une fonction de l'inclinaison, les lignes
qui representent l'un et l'autre de ces phénomènes, devraient
conserver une marche toujours parallèle. On peut en tirer la
conséquence interessante, que la position des deux poles magnétiques n'est pas la seule qui règle les phénomènes de l'inclinaison et de la declinaison dans les différentes parties du
globe ; mais qu'il existe encore une cause secondaire qui n'af-
fectant toutefois que tres faiblement la declinaison et l'inclinaison, et la dernière d'autant moins qu'on l'observe plus près
de l'équateur, exerce cependant sur les loix de l'intensité une
influence si puissante qu'elle en efface presque tous les carac-
tères déduits par la théorie."

 M. Erman's conclusions, in respect to the non-parallelism of
the lines of dip and intensity, and the insufficiency of a single
magnetic axis to represent his observations, were almost identical with mine. Our difference, in regard to the particular
class of the phænomena which were most at variance with that
hypothesis, arose from the different parts of the globe which
had been the field of our respective researches.

 I have next to state the inferences of M. Hansteen as *an
experimentalist*, drawn from his observations in his own extensive journeys. This need occupy the less space, because I
have already† endeavoured to show, as clearly as the necessity

　* *Mémoires de l'Acad. Imp. des Sciences de St. Petersburg*, 1831, (*Bulletin
Scientifique*).
　† Fifth Report of the British Association, p. 72—73.

of great condensation would admit, the arrangement of the lines of intensity, and their systematic departure from parallelism with those of the dip, which, in his theory of four poles, founded on the assemblage and study of the earlier observations of the dip and variation, M. Hansteen had anticipated, previous to his own experiments. It is sufficient to show, as may be done by a single sentence written since his return from Siberia, that the results of these have accorded with his previous views. "Thus is confirmed in the clearest and most satisfactory manner what I had earlier inferred from the two other magnetic phænomena ; namely, that in the northern hemisphere there are two magnetic centres, or poles ; and that the westernmost, in North America, has a sensibly greater intensity than the easternmost in Siberia*."

Having thus shown the concurrent opinions which those who have most extensively engaged in the experimental inquiry have been led to form, it remains to place the facts themselves in a convenient manner before the general reader. The complete view of the systematic difference in the course of the two kinds of lines is best obtained, by comparing the map of the intensity lines in this Report with M. Hansteen's map of the dip lines for 1780, in the Fifth Report of the British Association†. The lines of dip have undergone some changes since that period, but none which much affect their general configuration. All readers, however, may not have that volume at hand, and I have therefore traced in Plate I. the course of the line of equal intensity which passes through our own islands, for 160 degrees of longitude, and have exhibited it in comparison with the neighbouring lines of dip. The line of intensity, shown by the continuous line, is taken from the general map accompanying this memoir. The portions of dip-lines, marked by the dotted lines, are taken from M. Erman's map drawn from his own observations, in the *Annalen der Physik*, vol. xxi. The intensity line, which in the meridians of 280° and 290° is in close juxtaposition with

* *Ann. der Physik*, vol. xxviii. p. 579.

† I may take this opportunity of stating that the sea portions of M. Hansteen's map of the dip in 1780 rest on the authority of between 900 and 1000 observations of the dip made at sea between the years 1767 and 1788, and that these are tabulated in the Appendix of the *Magn. der Erde*. The observation of the dip at sea in favourable weather was the habitual practice of many of the scientific navigators of that period, such as Le Gentil, La Perouse, Ekeberg, Lewenhorn, and our own countrymen Phipps, Hutchins, Abercrombie, and Pickersgill. It is much to be wished that it were a more frequent practice now. M. Erman, in his voyage from Kamtschatka to Europe, found a number of days sufficiently favourable to enable him to observe the dip in not less than 167 geographical positions at sea.

that of 50° of dip, successively intersects in its eastern progress all the lines of dip between 52° and 73°, with which latter it coincides in lat. 60° and long. 10°; it then again descends, intersecting successively, a second time, the same lines of dip, until it touches that of 57° in long. 70°. When it is seen that the *same* line of intensity successively coincides with the lines of dip of *twenty different degrees*, it must be admitted that their systems are not parallel, and that the conclusion was justly drawn, that the facts could not be represented by an hypothesis in which the intensity should vary as any function of the dip. A conclusion by no means at variance, however, as has been erroneously imagined, with their having a causal connexion.

Nor is the fact of non-parallelism confined to the northern hemisphere; on the contrary, the southern hemisphere exemplifies it in a still more striking degree.   Thus we have in South America the line of unity under a dip of 0, as observed by M. de Humboldt in Peru; and at the Cape of Good Hope, the same line of unity under a dip *exceeding* 50°, as shown by the concurrent observations of Captains de Freycinet and Fitz Roy; whilst at Port Desire and at the Falkland Islands, these officers found an intensity of 1·36, with nearly the same dip as had been found at the Cape of Good Hope accompanying an intensity less than unity.

In M. Erman's dip-lines (Plate I.), which represent his own recent observations, and are quite independent of pre-existing evidence, we see the same double flexure, of which the importance, in its bearing on physical causes as well as on empirical laws, was pointed out in the Fifth Report of the British Association, page 67.   This double flexure takes place also in the intensity lines, but in a more marked degree. In both series of lines the radii vectores drawn from the geographical pole have two maxima and two minima; a line joining the parts of each curve which approach nearest to one another, i.e. at the points of minima, will divide the area into two unequal portions, the larger comprehending the American, and the smaller the Siberian centre of attraction.   But there is a distinction in this respect between the two series of curves of dip and intensity, which has been pointed out by M. Erman, and is illustrated by the annexed diagram (Plate II.), taken from his paper in the *Annalen der Physik*, vol. xxi. The diagram represents the northern hemisphere, on which the curves of intensity of 1·45 and of 75° of dip are drawn. The longitudes of the maxima of both these curves are nearly the same; but not so those of the minima. In the curve of dip, the minima

are in the longitude of 35° and 140°; in the curve of intensity
in those of 20° and 175°. The Siberian portion of the intensity curve bears consequently a larger proportion to the whole
area of that curve, than the Siberian portion of the dip-curve
does to its total area. From the general resemblance of the
several lines of dip to each other, and of the several lines of
intensity to each other,—the characteristics of each being
always marked, though gradually softening as they approach
the middle regions of the globe,—the features of distinction
which are thus strongly marked in the curves compared by M.
Erman, must exist also in a greater or less degree in many.
Here, then, is another striking and systematic difference in the
two species of magnetic lines*.

*2. The lines of intensity in the northern hemisphere system-
atically indicate the existence of two centres of attraction of
unequal force.*
The examination of the graphical representation of these
lines in the maps will convey a clearer apprehension of this
systematic indication than a lengthened verbal description.
The higher the values of the intensity of each isodynamic line,
—in other words, the nearer the lines approach the centres of
attraction,—the more unequivocal is their testimony. The
smaller areas included by the curves in the Siberian quarter
mark the less extensive influence and inferior power of the
Siberian centre. Looking next at the values of the intensities
represented by the lines, we find in the neighbourhood of New
York, a portion of a line of 1·8, to which there is no equivalent in Asia. The highest intensity there is 1·76, observed by
Lieut Due at Viluisk, which M. Hansteen believes, and with
great probability, derived from the configuration of the lines, to
be the highest existing in that quarter. It is improbable,
moreover, that the greatest intensity in the American quarter
should be found so far south as New York; the configuration
of the lines, as shown particularly in the north polar map, in-
dicates the maximum to be nearer Hudson's Bay†.

* M. Erman remarks that the difference is of that character which would
appear to indicate for the Asiatic centre a less depth beneath the surface than
the American.
† Since the above was written, the first number has reached London of the
*Observations Météorologiques et Magnétiques faites dans l'étendue de l'Empire
de Russie,* which have been confided to the editorship of M. Kupffer. In the
introduction we have a formal recognition of the existence of the Siberian pole.
" La Russie est aussi la terre classique du magnétisme terrestre. Il y a un
pole magnétique dans le nord de la Siberie."

3. *The two centres of magnetic attraction in the northern hemisphere are not at opposite points; in other words, the difference of geographical longitude between them is not 180°, measured both ways.*

This is also best evidenced by inspection. Their distances apart are more nearly 200° measured across Greenland and Norway, and 160° across Behring's Strait.

4. *The magnetic intensity is unsymmetrically distributed in the meridians of the northern hemisphere.*

This is a consequence of the two centres being nearer to each other in the one direction than in the other. If we imagine the hemisphere to be divided into two equal sections, by a plane coinciding with the meridians of 100° and 280° (Plate V.), the American division, which we may call the western section, will contain both centres of attraction, and a higher measure of intensity will be seen to be spread over its meridians than in the corresponding latitudes in the eastern section. Thus we find, that in 150 meridians, or in five-sixths of the eastern section, no intensity of so high a value as 1·7 has been found within the range of observation, and probably does not exist; whilst in the western section there is not a single meridian in which a higher intensity than 1·7 is not found. Europe is situated nearly midway between the centres at their widest separation, and we find that throughout Europe (with possibly the exception of its S.W. extremity in Spain), the magnetic intensity is weaker in every latitude than in the same parallels elsewhere in any other part of the hemisphere.

5. *The lines of intensity in the southern hemisphere have a general analogy with those in the northern hemisphere.*

The materials from whence conclusions may be drawn are fewer in the southern than in the northern hemisphere; but aided by our acquaintance with the magnetic system and distribution in the latter, we are enabled to trace the general analogy of the two hemispheres, though the particular conclusions in the case of the southern must necessarily be less determinate and exact than those we have hitherto discussed.

We have already seen that the lines of dip and force depart from parallelism with each other even more in this hemisphere than in the northern. We may also perceive in the portions of the curves, which observations have as yet enabled us to trace, evidence of the same double flexure which in the other hemisphere we have seen to be characteristic of two centres of governing influence. The radii vectores carried from the south geographical pole would have also two maxima and two minima. The New Holland curves inclose larger areas than the South American, indicating that the centre to which they more especially belong is more powerful than the other. We have another indication of the same fact in the appearance in Van Diemen's Land of an intensity exceeding 1·8, which in the other hemisphere we have seen to characterise distinctively the centre of primary influence. The coincidence in this respect in the two hemispheres is very striking; not only is the highest intensity yet observed in the one, (1·80 at New York,) matched by the nearly identical value of 1·82 at Hobart Town, but the geographical latitudes of the two observations are also nearly identical, New York being in 40° 43′ N. and Hobart Town in 42° 53′ S.; both being unexpectedly low latitudes in which to find such high intensities.

With regard to the geographical positions of the centres in the southern hemisphere, the observations are yet too few and too distant from them to admit of their localities being assigned with any fair degree of approximation; but by comparing the observations in Southern Africa, and on the east coast of South America, with those of the corresponding parallels in the better known hemisphere, we are able to infer with considerable probability, that the southern centres are not only not in opposite points of the hemisphere,—that is to say, distant 180 degrees of longitude from each other, measured both ways,—but that they are nearer to each other in the one direction, and more distant in the other, than is the case with the centres of the northern hemisphere. We have seen that in the meridians of Europe, where the northern centres are widest apart, the lower intensities extend greatly northward, occupying latitudes which in all other parts of the hemisphere possess a higher intensity. In the southern the same thing takes place, but in greater degree. The line of unity, once thought to be the minimum intensity on the globe, is found on either side the Atlantic in south latitudes exceeding 30°; whence we may conclude that in the higher latitudes of the southern Atlantic, a much lower intensity prevails generally than the lowest intensities in the same latitudes in the northern hemisphere; evidencing that the space between the influential centres is wider in that quarter of the southern, than in the corresponding quarter of the northern hemisphere.

The converse of this should be found in the Pacific section. As the southerly inflection of the lines of low intensity in the South Atlantic is the greatest, so should their southerly inflexion in the opposite section of the hemisphere be the least, of the inflections which these lines undergo in either hemisphere. The observations by which this inference might be confirmed are few, but none give a contrary indication. Every observation in the South Pacific section shows that a higher intensity prevails there than in equal latitudes in the North Pacific section; and, as far as the lines can yet be traced from the observations, the inflection in the South Pacific does appear to be the least marked in character, and to extend over the fewest meridians. It is of course the lines of higher intensity which would-afford the more decisive evidence, because their characteristics are more marked; but the authorities for these are few in the part of the space between New Zealand and South America, where they could most illustrate the point in question.

In review, we conclude, therefore, that, as far as observations have yet been made in the southern hemisphere, they accord with a system analogous to that in the northern, of two centres, of unequal force, and at unequal distances apart. The observations further render it probable, that the distances between the centres are still more unequal in the southern than in the northern hemisphere. Admitting the small difference of distribution from this cause, there does not appear reason to suppose that there is any general inequality in the magnetic charge of the two hemispheres; on the contrary, there is every appearance that they have the same.

6. *If the globe be divided into an eastern and a western hemisphere by a plane, coinciding with the meridians of* 100° *and* 280°, *the western hemisphere, or that comprising the Americas and the Pacific Ocean, has a much higher magnetic intensity distributed generally over its surface, than the eastern hemisphere, containing Europe and Africa and the adjacent part of the Atlantic Ocean.*

This is a corollary from (4) and (5) rather than a distinct proposition. The four centres being in the western hemisphere a higher intensity will prevail generally in its meridians; and this is accordant with the whole body of observations distributed over the globe (Plate V).

The equality of the magnetic charge in the northern and southern hemispheres and its inequality in the eastern and western, are important features of the magnetic system manifested by the observations of intensity.

7. *The distribution of the intensity in the intertropical regions is accordant with the conclusions already drawn, of two governing centres in each hemisphere.*

As the lines of higher intensity are those which have the

characteristics of the system most strongly marked, I have chiefly employed them, where observations would permit, in describing its general features. The characteristics soften gradually as the distance increases from the governing centres; but even in the intertropical regions the distribution of the intensity and the arrangement of the lines contribute their testimony to the same system. I have nowhere attempted to assign the precise geographical positions of the centres; and in regard to those of the southern hemisphere especially, have expressly stated, that the facts yet acquired would not enable this to be done within fair limits of approximation. Thus much, however, may be safely said in regard to them, that the primary in the southern, and the secondary in the northern, are at the present time not far from the same meridian; and that the primary in the northern, and the secondary in the southern, are similarly situated, except that their difference of longitude is somewhat greater. If we respectively connect the centres, which thus approximate in longitude, by lines on the globe crossing the equator, the lines will mark those localities within the tropics where the influence of the centres should produce a higher intensity than elsewhere in the same latitudes. Thus we should have two maxima in the intertropical regions; and these should not be in opposite meridians, because the centres are unsymmetrical. Such is actually the distribution of the intensity in these regions. The isodynamic lines which represent unity are the weakest which run unbroken round the globe, and appear twice in every meridian; these approach each other in the meridians of 110° and 260°, whilst, intermediately, they recede from each other, and inclose spaces occupied by a still weaker intensity; the largest of these spaces, corresponding to the widest interval between the centres, is of 210 degrees of longitude, and the smallest of 150 degrees. In the middle of the largest, as the point most distant from all the four centres, we should expect to find the weakest intensity existing anywhere at the surface of the globe; and accordingly at St. Helena, which is nearly in that situation, the intensity observed by Captain Fitz Roy, 0·84, is the lowest determination recorded in this report, and is the locality of the weakest intensity yet observed on the globe. Between St. Helena and the lines of unity on either side, we should have a line representing the value of 0·9, a part of which has been extremely well determined by concurrent observations. This line, being comprehended by the lines of unity, is necessarily a closed one. Observations are yet wanting to show whether the intensity descends as low as 0·8 in the middle of the larger space, or as 0·9 in the smaller space, which has its locality in the Pacific*.

We may also trace in the intertropical regions another consequence of the inequality of force of the primary and secondary centres. Where the lines of unity approach each other in the Pacific, the primary is to the north, the secondary to the south; the latitude in which the lines approach is consequently to the south of the equator. In the Indian Sea the primary is to the south, and the secondary to the north; and here the latitude in which the lines of unity approach each other is to the north of the equator.

Every geographical meridian has a point of minimum intensity; if these points in different meridians were connected by a line, that line would separate the intensities of the northern from those of the southern magnetic hemisphere. It would be in some respects analogous to the line of no dip, but it would not be a line of equal intensity, as it would consist of intensities varying from unity to the lowest on the globe. Such a line traced on the map is found to differ very considerably in geographical position from the line of no dip.

8. *The geographical position of the maximum of intensity in the North American quarter is not the same with that of the maximum of dip, or with that of the point of convergence of the variation lines.*

It will be necessary here to enter into rather more precise geographical positions than we have hitherto done. In regard to the maximum of dip we cannot err widely in taking the latitude and longitude where Capt. James Ross observed the dip of 89° 59′ in 1831, viz. 70° N. and 263° E. That this is also very nearly the spot to which the variation lines converge may be shown abundantly by the observations made in the different polar voyages by sea and land[†]. It is marked by an asterisk

---

* Since the above was written Mr. Erman's sea observations have been received; he crossed the space in the Atlantic included by the line of 0·9 some degrees to the west of St. Helena, and, midway between the north and south portions of that line, found the intensity diminished below 0·8. Captain Fitz Roy's observation at St. Helena is consequently no longer the lowest observed on the globe; and it is probable that even a lower intensity than was observed by M. Erman would be found a few degrees to the south of St. Helena, and nearly in the meridian of that island.

† M. Hansteen, who has brought together the observations of dip and variation made in the different polar voyages, finds that the variations observed to the north of the latitude in which the dip is 90° and in the vicinity of that dip, converge to a point a little to the north of that latitude; and conversely, that the variations observed to the south converge to a point south of that latitude; or, more exactly, that the curves of highest dip are ellipses, having their greater axes

in the North Polar map annexed to this report. If the reader
will now refer to that map (Plate IV.), he will see that this position
will by no means accord with that which the observations point
out for the maximum of intensity. We are not, indeed, enabled
to assign the position of the latter as nearly as in the case of
the dip ; but it must clearly be in a much lower latitude. The
intensities observed in Baffin's Bay and the Polar Sea have
all a much lower value than at New York ; and the general
configuration of the lines of intensity would rather point to a
maximum in the vicinity of the shores of Hudson's Bay.

This remarkable feature of the system was first brought to
notice in the account of my magnetic observations published
in 1825[*]. In a point of so much interest, it is natural to in-
quire whether there is any indication of a similar separation at
the principal pole of the opposite hemisphere. Observations
as yet do not enable us to assign with sufficient approximation the places of the maxima in that quarter ; but we are in
possession of a leading fact, which, by its complete analogy
with the phænomena at New York, gives strong ground for
believing that in the southern hemisphere also the places of
the maxima of the two phænomena are distinct. I have already
noticed the almost identity of the force at Hobart Town and
New York, under nearly equal geographical latitudes; but there
is yet another feature which completes the analogy, and bears
directly on the point now treated of. At New York we have
the highest intensity of the northern hemisphere, 1·80, *with a
dip of* 73° 07' ; at Hobart Town the highest intensity of the
southern hemisphere, 1·82, *with a dip of* 70° 35'. In both hemispheres *the highest intensity united with a comparatively low
dip.* Nor in that quarter is Hobart Town a solitary instance of

in a north-west and south-east direction, and that the variation lines converge
not to the point of 90° but to points in this axis. Small differences of position,
however, have no effect on the reasoning in the text.

[*] It has been viewed by M. Kupffer as having a direct and important bearing on the very interesting question of the physical nature of the magnetism of
the earth. In the *Ann. der Physik*, vol. xv., after describing the course of the
isogeothermal lines (or lines of equal temperature of the earth at 25 metres below its surface) between the meridians of 80° west and 60° east of Paris, he has
discussed the influence which the facts represented by those lines should
have on the magnetic dip and force, in the case of the earth's magnetism being
superficial and induced. The differences of surface temperature affecting the
intensity but not the dip would cause the isoclinal and isodynamic lines to separate where otherwise they might have been accordant; and would especially
separate the places of the maxima, causing the maximum of intensity to be in
the lower latitude. M. Kupffer considers the fact of their being thus separated
as giving probability to the aforesaid view of the physical nature of the earth's
magnetism.

high intensity with comparatively low dip; at King George's Sound and Sydney, in 34° and 35° south latitude, Captain Fitz Roy found intensities of 1·71 and 1·68 with dips of 64° 41' and 62° 29'.

Should such a separation exist at the secondary centres, it cannot be expected to be of so striking a character. I wish not to anticipate the more able discussion which we may expect on this point from M. Hansteen, whose long and arduous journeys were undertaken expressly to determine with exactness all the phænomena of the Siberian pole. I will confine myself, therefore, to noticing his remark already referred to, that he believes the intensity observed at Viluisk to be the highest intensity existing in Siberia. Should this be so, the highest intensity in that quarter is certainly not in the same locality as the highest dip*.

Our knowledge of the phænomena in the neighbourhood of the secondary centre in the southern hemisphere is not sufficient to throw any light on this question.

With regard to the direction which the lines of higher intensity may be conceived to take around their maxima in the northern hemisphere, we should infer from the observations that the line representing 1·8 must be a closed curve around the North American maximum only; as must also be that of 1·9, supposing such to exist.

The North American portion of the line of 1·7 appears also to be nearly, if not quite, a closed curve. Encompassed on the north, east, and south, by intensities of less value, the western is the only direction open for its connection with the Siberian portion of the same line. The situation of the two branches of the line of 1·7 in the west of America is marked by the observations;—the southernmost crossing the lower waters of the Columbia River,—and the northernmost between Sitka and Melville Island. Whether these branches join and form a closed curve, or whether they communicate with the Asiatic portion of the same line in some such courses as is represented by the dotted line in the polar map, observations do not yet enable us to decide. No intensity of so high a value as 1·7 has yet been observed between Sitka in 224°, and the meridian

---

* It is much to be desired that the observations in Siberia should be still further completed by a series of determinations along the shores of the polar sea. If the view here taken be correct, these should exhibit higher dips and lower intensities than were observed at Viluisk. From the liberal support which the Russian government gives to the prosecution of magnetic inquiries we may expect that such observations will not be long wanting.

of 138° in Siberia; and it is possible that a navigator sailing from the Pacific through Behring's Strait, and passing the Bay of St. Lawrence where Admiral Lütke observed 1·65, might proceed to the northward having the spaces included by the closed curves of 1·7 on either side of him.

The space inclosed by the curve of 1·8 possesses a very high degree of magnetic interest, and is well deserving of being traversed by observations as frequent and as accurate as those of MM. Hansteen and Erman in Siberia. The greater part of it is in the British dominion, and over a considerable portion at least convenient means of locomotion are to be found. The British Association had but to express the wish that a magnetic survey of the British Islands should be made, and it was at once responded to by some of its own members. The present volume contains the record of the completion of that undertaking; and it may be permitted to one of the contributors to that work to express a hope, that the attention of the Association may now be given to the British possessions abroad. In the extensive territory under British dominion in India, not a single determination has yet, I believe, been made of the magnetic intensity, and but few of either of the other phænomena. From the well-known zeal of the officers of the Indian service, a recommendation in the proper quarter would speedily cover that large portion of the earth's surface with accurate magnetic determinations. But the Canadian quarter is of prominent interest; a correct delineation of the lines of variation, dip, and intensity in the space included by the curve of 1·8, or in even a portion of that space, would have a high value in theoretical respects. The accomplishment of this service is not altogether beyond the compass of individual means, and needs not, like a southern voyage, await the success of an application to Government. It requires only for its proper execution, that it should be the principal object of the person undertaking it, and that he should be provided with adequate instruments. Were the wishes of the Association expressed in regard to Canada, as they were in regard to the British Islands, I have little doubt that they would soon be complied with by members of their own body*.

---

* The ground which Capt. Back traversed in his journey in search of Capt. Ross in 1833 and 1834 is of great interest as regards the magnetic intensity; and had that officer been furnished with suitable instruments, and had it accorded with his other objects to have made observations in the manner of MM. Hansteen and Erman at every halting-place, his results might have possessed great value.

The vibrations of the dipping-needle, which he employed to measure the in-

9. *The highest intensity already observed is more than twice as great as the lowest.*

The intensities observed at New York and Hobart Town, compared with that at St. Helena, are as 1·81 to 0·84, or as 2·16 to 1.

St. Helena is not the lowest intensity; and the force at New York and Hobart Town cannot be viewed as abso-

tensity, appear to have been subject to a considerable instrumental uncertainty ; and the needle lost magnetism during the absence from England to a large amount, but at what time the loss took place is not very obvious from the observations. Under these circumstances I have not felt that I could assign with sufficient confidence the value of the intensity relatively to Europe at any of Capt. Back's American stations. By grouping them, however, and comparing the values of the intensity in different groups, relatively to each other only, and not relatively to Europe, we may considerably lessen the effect of the irregularities above mentioned, and obtain an indication, which, if we could view it as sufficiently clear from instrumental uncertainty, would possess much interest. For example, if we group neighbouring stations as in the subjoined table, and make the intensity at New York the unity of the comparison, we have as follows : viz.

Station.	Date.	Lat. North	Long. West	Time of Vib.	Therm.	Mean				Intensity.
						Lat.	Long.	Time of Vibr.	Ther.	
New York............	1833 Apr.	40 42	74 01	s. 1·2857	68·	40 42	74 01	s. 1·2857	° 69	1·000
Fort Alexander ...	Jun.	50 37	96 21	1·2432	70·5					
Cumberland House	July	53 58	102 22	1·2643	59·5	53 20	102 13	1·2631	68	1·027 (a)
Isle à la Crosse	July	55 25	107 55	1·2969	78·5					
Fort Chipewyan ...	July	58 42	111 19	1·3000	95·					
Fort Resolution ...	Aug.	61 10	113 45	1·2387	65·6	59 56	112 32	1·2693	80	1·031 (b)
Fort Reliance ... {	Oct.	62 46	109 01	1·2750	44·					
	1834 May	......	......	1·2844	49·	62 46	109 01	1·2792	40	0·997
	Oct.	......	......	1·2781	28·					
Musk Ox Rapid ...	July	64 41	108 08	1·2873	64·					
Rock Rapid	July	65 54	98 10	1·2800	87·					
Point Beaufort......	July	67 41	95 02	1·2975	72·	66 51	98 19	1·2838	70	1·002 (c)
Montreal Island ...	Aug.	67 47	95 18	1·2885	74·					
Point Ogle	Aug.	68 14	94 58	1·2656	53·					

Here we see that the groups (a) and (b), which have their mean position about 53° N. and 102° W., (258 east), and 60° N., and 112½ W. (247½ east), have a higher intensity than the more northern group (c), which has its mean position about 67° N. and 98° W. (262 east). These groups (a) and (b) have also a higher intensity than that of Fort Reliance to the north, or New York to the south. New York, Fort Reliance, and the northern group (c), scarcely differ in the values of their respective intensities. This arrangement is quite conformable with the lines in the polar map.

I have taken Capt. Back's observations from Mr. Christie's paper in the Phil. Trans. for 1836 ; the times of vibration at the stations in America being those contained in the table page 393. That table shows that the needle was vibrated at

lutely the highest.   If we suppose the minimum to reach
0·74, (one of M. Erman's sea observations is 0·743) and the

every station with its face to the face of the instrument, and that at *some* of the
stations it was also vibrated in the reverse position.   Where this has been done
there often appears a considerable difference between the times of vibration
at the same place in the two positions, which must be ascribed to instrumental
defect.   It does not appear to have been of the nature of a constant error in
either position of the needle, as sometimes one position gives the highest intensity and sometimes the other.   I have taken the twelfth column just as it stands,—
that is, the times of vibration in the position which was everywhere observed,
as there can be no question of the comparability of those with each other ; and
I have reduced the times of vibration to an uniform temperature by the coefficient which Mr. Christie found for that needle ; but I have introduced no other
corrections, either for loss of magnetism or on any other account.   I have grouped
the results by taking the mean latitude, longitude, and intensity of the neighbouring stations, connected by brackets.

If the intensities are taken from a mean of all the observations at each of the
stations, including those in the reversed, as well as in the direct position of the
needle, the inferences drawn above are somewhat strengthened, as is shown in
the following table :—

Station.	Lat. North.	Long. East.	Time of Vib.	Ther.	Intensity.
New York	40 42	285 59	1·2857	69	1·000
Group (a)	53 20	257 47	1·2644	69	1·033
Group (b)	59 56	247 28	1·2607	80	1·045
Fort Reliance.....	62 46	250 59	1·2758	40	1·002
Group (c)	66 51	261 41	1·2857	70	0·999

Mr. Christie, in combining the observations at different stations and in different positions of the needle, has followed a somewhat different course, and has
arrived at somewhat different conclusions.   With more perfect instruments,—
with observations alike complete at all the stations,—and repeated at New
York as well as in London, to test the permanency of the needle's magnetism,—
there would not have been room for any difference of view.   The only result
absolutely deducible from the observations, and in which all persons must
agree, is the comparability of the intensities at the different stations of the
northern group with each other, and with Fort Reliance ; as the observations of May. and October, 1834, show by their agreement that during
that interval the needle underwent no change.   The conclusion to be drawn
from this portion of the observations, which are as strictly comparable as
the imperfection of the instrument permits, is, that in the district which it
comprises no consistent alteration takes place in the intensity.   If any small
alteration does take place, it would require a more delicate instrument than
Capt. Back was furnished with to determine it.

It is in these countries that the statical method of Professor Lloyd would be
of the greatest advantage.   I have already had occasion to speak of the disadvantage to which the method by horizontal vibrations is exposed in countries
of very high dip, where every error in the dip is magnified to a high degree in
its effect on the intensity deduced ; and of the preference due in such cases to
the vibrations of a dipping-needle.   But it is well known that this latter method,
though a trust-worthy, is far from being a delicate test of differences of mag-

maximum 1·85, the proportion would be 2·5 to 1. It seems probable that this is rather under than over the difference existing in the present distribution of the intensity. If the centres change their relative places, by having unequal motions, both the absolute and the relative values of the maximum and minimum must be variable.

This report has already occupied so large a portion of the annual volume, that I feel the propriety of not permitting the inferences of an individual judgment to trespass further on its pages. I have endeavoured, to the best of my power, to place the facts themselves before the reader in such a manner, that, on the one hand, he may have no difficulty in tracing every observation to its original source,—and on the other, that by the assemblage of the results in one view, he may be enabled with the greater facility to draw his own conclusions.

Having in a former report described M. Hansteen's theory of the magnetism of the earth, and given the formulæ for the variation, dip, and intensity deduced from his hypothesis of two excentric axes of unequal force, it may be expected that I should conclude this report by comparing some of the observed intensities with the results computed by the formula. I may therefore add a few words to show that the proper time for a detailed comparison of this kind has not yet arrived, because observation is still in arrear of theory. Until observation has supplied the materials which theory has required for the correct assignment of the elements of calculation, such a comparison could not be otherwise than imperfect.

The geographical positions of the magnetic poles in the *Magnetismus der Erde* were derived from observations made between 1787 and 1800, which were insufficient to furnish them in more than a very general manner. Since that period also, changes, of the nature anticipated by M. Hansteen, appear to have taken place in the positions of the poles; which consequently require to be assigned afresh (as well as more correctly), in order that the results computed by the formula may represent observations of a more recent date. The materials proper for this purpose are observations in the vicinity of the

netic intensity, even with a good instrument, on account of the shortness of the period during which the needle will continue to vibrate, and the consequent necessity of commencing with a large arc of vibration. With an inferior instrument the limits of error are of course much wider still. In high magnetic latitudes the statical method deserves a decided preference over the method of *horizontal* vibrations, inasmuch as a moderate error of the dip will scarcely have an appreciable effect on the intensity ; and over that by *vertical* vibrations, inasmuch as it admits of much greater exactness.

magnetic poles themselves. In the northern hemisphere, these are far more ample and exact than at any former period, owing in great measure to the interest excited by the publication of M. Hansteen's theory. But the corresponding observations in the southern hemisphere are yet wanting; and until these are supplied, we cannot advance beyond an anticipation, more or less confident, of the eventual accordance of the hypothesis, when the correct elements of calculation shall have been obtained; and in this view, we may at least say thus much in regard to the general accordance of the hypothesis with the observations of intensity, that if we omit the consideration of the higher latitudes, where the contemporaneous and correct positions of the magnetic poles are most essential, the formula, even with the elements derived from the earlier and less perfect observations, both represents all the leading features of the system, and shows a fair approximation in individual cases.

The method in which this science has progressively advanced is strikingly illustrative of a passage in Mr. Playfair's writings, in which the distinct offices of theory and experiment, and the value of their co-operation in inductive investigation, are well described. " In physical inquiries the work of theory and observation must go hand in hand, and ought to be carried on at the same time, more especially if the matter is very complicated, for then the clew of theory is necessary to direct the observer. Though a man may begin to observe without any hypothesis, he cannot continue long without seeing some general conclusion arise; and to the nascent theory it is his business to attend, because by seeking either to verify or to disprove it, he is led to new experiments and new observations. He is led also to the very experiments and observations that are of the greatest importance; namely, to those *instanciæ crucis* that naturally present themselves for the test of every hypothesis. By the correction of his first opinion a new approximation is made to the truth, and by the repetition of the same process certainty is finally obtained. Thus theory and observation mutually assist one another; and the spirit of system, against which there are so many and so just complaints, appears nevertheless as the animating principle of inductive investigation. The business of sound philosophy is, not to extinguish this spirit, but to restrain and direct its efforts. It is therefore hurtful to the progress of physical science to represent theory and observation as standing opposed to one another."

The earlier observations of terrestrial magnetism were made without reference to theory. As facts accumulated general conclusions arose. Their elaborate examination conducted to an hypothesis of four magnetic poles; and this, to the suggestion of new experiments to verify or disprove it.   In the northern hemisphere the verification is complete, affording signal proof of the value of experiment directed by theory.   A similar verification in the southern hemisphere is yet wanting; and the observations necessary for that purpose will also supply those elements of calculation whereby the hypothesis may be fitted for a detailed comparison with facts.   This will be the next "step in the advancement of knowledge;"—the next "term of a series that must end whenever the real laws of nature are discovered";—but which, in its progression, fitly prepares the way for their discovery.

I have already adverted to what the influence of the Association may effect, in causing the spaces yet vacant on the map, in the British possessions in India and Canada, to be filled. But beyond all comparison, the most important service of this kind, which this or any other country could render to this branch of science, would be by filling the void still existing in the southern hemisphere, and particularly in the vicinity of those parts of that hemisphere which are of principal magnetic interest.   This can only be accomplished by a naval voyage; for which it is natural that other countries should look to England. That the nations that have made exertions in the same cause do look to England for it, cannot be better shown than by the following extract of a letter of M. Hansteen's, which I take the liberty of introducing here, both for this purpose, and because it expresses in so pleasing a manner, the praise that is so justly due to his own country, and which I am sure will be cordially responded to by all who cultivate science in this country, and particularly by those who know the kindly feeling with which Englishmen are ever welcomed in Norway.

   " C'est le Storthing (la Chambre des Députés) de la Norvège, qui a donné les frais à l'expédition en Sibérie. On a fait cela dans un tems où on a refusé les dépenses pour un château de résidence pour sa Majesté à Christiania. Dans un tems, où une telle économie a été nécessaire, il est très honorable, qu'une Chambre, composée de toutes les classes du peuple, même d'un grand nombre de paysans, a *unanimement* résolu de donner les frais pour une expédition purement scientifique, dont les résultats n'auront jamais aucune utilité économique pour la patrie, et dont on ne comprenait pas la haute valeure scientifique.   Regardé les ressources très-bornés de notre pays, c'est une générosité presque sans exemple.

   " Comme la petite Norvège a fourni toutes les observations entre les méridiens de Greenwich et de Ochozk, et entre les parallèles de 40° et 75° de latitude boreale, il ne me semble pas une demande trop grande ou immodeste à l'Angleterre, si grande, si riche, si puissante, qui a nécessairement un plus grand intérêt dans toutes les sciences combinées avec la navigation, de fournir toute la partie méridionale de la carte. Une telle entreprise doit réfléchir une splendeur à la nation, et payera à la fin les frais par des résultats aussi utiles pour les sciences que pour la navigation. Il ne faut plus dans notre tems laisser l'avancement des sciences au hasard. Par des observations fragmentaires et discontinués on a tâché avec grande peine d'étudier les phénomènes magnétiques de la terre pendant deux ou trois siècles. Par deux ou trois expéditions litéraires, arrangées *exprès pour ce but*, on pourrait en peu d'années avoir une collection plus complète, et d'une plus grande utilité pour la théorie."

The subject has in every way a claim on this country. The existence of four governing centres, and the system of the phænomena in correspondence therewith, was originally a British discovery. The sagacity of our countryman Halley was the first to penetrate through the complexity of the phænomena, and to discern what is now becoming generally recognised. England was also the first country which sent an expedition expressly for magnetic observation, namely, that of Halley in 1698 and 1699. Whilst approving and cordially co-operating in magnetic inquiries of other kinds which have their origin in other countries, it is right that we should feel a peculiar interest in that in which we have ourselves led the way, especially when its object is subordinate to none.

As the research would require to be prosecuted in the high latitudes, a familiarity with the navigation of such latitudes would be important in the person who should undertake this service; and a strong individual interest in the subject itself would be of course a most valuable qualification. I need scarcely say that the country possesses a naval officer in whom these qualifications unite in a remarkable degree with all others that are requisite; and if fitting instruments make fitting times, none surely can be better than the present.

Viewed in itself and in its various relations, the magnetism of the earth cannot be counted less than one of the most important branches of the physical history of the planet we inhabit; and we may feel quite assured, that the completion of our knowledge of its distribution on the surface of the earth, would be regarded by our cotemporaries and by posterity as a fitting enterprise of a maritime people; and a worthy achievement of a nation which has ever sought to rank foremost in every arduous and honourable undertaking.

ERRATA.

Page 32, line 4, *omit* giving each equation a weight proportioned to the
number of observations which it represents.

„ 8, *for* δ = 68° 42′; r = — 0·013608, *read* δ = 68° 33′;
r = — 0·01405, equivalent to 71 geographical miles for
one degree of dip.

General Table.

Page. 44	Frazer's Lake	Intensity	*for* 1·724, *read* 1·734.
„	Stuart's Lake	„	*for* 1·736, *read* 1·745.
„	Fort Alexandria	„	*for* 1·710, *read* 1·714.
Page 45.	Multnomah River	„	*for* 1·669, *read* 1·660.
„	Sandiam River	„	*for* 1·683, *read* 1·672.
„	Columbia Rapids	„	*for* 1·679, *read* 1·671.
„	Thompson's River	„	*for* 1·710, *read* 1·701.
„	Oakanagan	„	*for* 1·707, *read* 1·701.
„	Wullawullah River	„	*for* 1·707, *read* 1·699.
Page 46.	St. Francisco	Longitude	*for* 235 45, *read* 237 35.
„	San Solano	„	*for* 235 36, *read* 237 36.
„	Monterey	„	*for* 236 00, *read* 238 00.
„	San José	„	*for* 236 00, *read* 238 00.
„	La Soledad	„	*for* 236 36, *read* 238 36.
Page 47.	San Antonio	„	*for* 236 42, *read* 238 42.
„	San Miguel	„	*for* 237 16, *read* 239 00.
„	St. Louis Obispo	„	*for* 237 20, *read* 239 20.
„	La Purissima	„	*for* 237 33, *read* 239 33.
„	Santa Ynez	„	*for* 237 49, *read* 239 49.
„	Santa Barbara	„	*for* 240 00, *read* 240 20.

And in the column of Intensities:

St. Francisco, Solano, *for* 1·610, *read* 1·614.
San José   *for* 1·605, *read* 1·607.
San Miguel   *for* 1·583, *read* 1·580.
San Obispo   *for* 1·583, *read* 1·580.
Santa Barbara   *for* 1·604, *read* 1·587.

www.ingramcontent.com/pod-product-compliance
Lightning Source LLC
Chambersburg PA
CBHW070840180526
45168CB00002B/903